Hydrogen and Fuel Cells: Innovations and Challenges

Hydrogen and Fuel Cells: Innovations and Challenges

Guest Editors

Hanqing Yang
Jiajia Yang
Zening Li
Zhengmao Li

MDPI

Basel • Beijing • Wuhan • Barcelona • Belgrade • Novi Sad • Cluj • Manchester

Guest Editors
Hanqing Yang
School of Automation Engineering
University of Electronic and Technology of China
Chengdu
China

Jiajia Yang
College of Science and Engineering
James Cook University
Townsville
Australia

Zening Li
College of Electrical and Power Engineering
Taiyuan University of Technology
Taiyuan
China

Zhengmao Li
Department of Electrical Engineering and Automation
Aalto University
Aalto
Finland

Editorial Office
MDPI AG
Grosspeteranlage 5
4052 Basel, Switzerland

This is a reprint of the Special Issue, published open access by the journal *Electronics* (ISSN 2079-9292), freely accessible at: www.mdpi.com/journal/electronics/special_issues/6YV90QCG9G.

For citation purposes, cite each article independently as indicated on the article page online and using the guide below:

Lastname, A.A.; Lastname, B.B. Article Title. *Journal Name* **Year**, *Volume Number*, Page Range.

ISBN 978-3-7258-3190-6 (Hbk)
ISBN 978-3-7258-3189-0 (PDF)
https://doi.org/10.3390/books978-3-7258-3189-0

Contents

electronics

MDPI

Article

Fuel Cell-Based Distributed Robust Optimal Scheduling for Combined Heat and Power Supply

Lei Xu [1], Yang Kou [1], Weile Liang [1], Alihan Bieerke [1], Yunshan Wang [1], Ji Li [1,*] and Zhi Yuan [2,*]

1 Electric Power Research Institute of State Grid Xinjiang Electric Power Co., Ltd., Urumqi 830011, China
2 Engineering Research Center of Renewable Energy Power Generation and Grid-Connected Control, Ministry of Education, Xinjiang University, Urumqi 830017, China
* Correspondence: lj24sg@163.com (J.L.); yz2019@xju.edu.cn (Z.Y.)

Abstract: At present, the safe operation of integrated energy systems is significantly affected by the considerable uncertainty inherent to wind and photovoltaic power generation. Based on this, this paper proposes an optimal scheduling model for integrated electricity, heat, and hydrogen-based energy systems on distributed robust optimization (DRO). Firstly, a combined heat and power microgrid system considering hydrogen energy systems was constructed based on the thermoelectric cogeneration characteristics of fuel cells and electrolyzers. Then, a data-driven two-stage distribution robust optimization scheduling model is built by combining typical historical data of wind power output, photovoltaic power output, and load. The results show that the distributed robust method reduces the running cost by 6% compared to the deterministic method. The proposed method and model are capable of meeting the demand for thermoelectric loads within the microgrid in a more cost-effective manner, thereby achieving stable and independent operation of the system.

Keywords: fuel cell; electrolyzer; distributed robustness; data-driven uncertainties

check for updates

Citation: Xu, L.; Kou, Y.; Liang, W.; Bieerke, A.; Wang, Y.; Li, J.; Yuan, Z. Fuel Cell-Based Distributed Robust Optimal Scheduling for Combined Heat and Power Supply. *Electronics* **2024**, *13*, 4172. https://doi.org/10.3390/electronics13214172

Academic Editor: Carlos Andrés García-Vázquez

Received: 8 September 2024
Revised: 18 October 2024
Accepted: 21 October 2024
Published: 24 October 2024

1. Introduction

The vigorous development of renewable energy and the construction of a low-carbon, clean, and efficient integrated energy system (IES) represent a crucial strategy for addressing the growing scarcity of fossil fuels and worsening environmental pollution [1–3]. The integrated energy system represents a crucial avenue for the low-carbon energy transition, facilitating the coordination and optimization of multiple energy sources to ensure a reliable and efficient energy supply [4]. The IES incorporates a substantial assortment of equipment [5], encompassing distributed renewable energy, cogeneration units, heat pumps, gas-fired boilers, distributed energy storage, and numerous other components. In recent years, the advent of new resources for hydrogen energy devices has prompted many experts and scholars to focus their attention on the optimization of integrated energy systems that incorporate these emerging devices. In [6], Fang et al. proposed a multi-timescale energy management solution for supplying electricity, hydrogen, and heat loads based on a multi-energy microgrid containing a hydrogen energy system. In [7], Song et al. proposed a coordinated optimization approach that considers both long-term and short-term hydrogen energy storage, demand response, and the influence of multiple uncertainties. They also developed a multi-timescale scheduling model for an integrated energy system based on zero-carbon hydrogen. To further explore the potential of multi-energy complementarity at numerous time scales under various operating conditions, in [8], Dong et al. proposed a refined modeling and co-configuration methodology for an electric–hydrogen–heat–gas integrated energy system (EHTG-IES) with a hybrid energy storage system (HESS). In [9], Ju et al. constructed a year-round hydrogen storage configuration model based on time-ordered operation simulation. This was performed to synergistically optimize the capacities of the electrolyzer, fuel cell, battery energy storage (BES), and hydrogen storage tank. This approach allows for the full play of the power and capacity complementarity of BES and generalized hydrogen storage, which can compensate for power and energy imbalance. In [10], Wang et al. developed a model of the electric–thermal coupling of a hydrogen storage unit and put forth an optimization plan for an integrated system with hydrogen storage that considers the use of thermal and electrical power, thereby lowering carbon emissions and raising the overall energy efficiency level. In [11], Shi et al. proposed a hybrid energy scheduling model for a multi-energy microgrid integrating a hydrogen energy storage system and a thermal storage system and verified the effectiveness of the proposed hybrid model for microgrid energy scheduling. In [12], Wang et al. proposed an integrated energy system operation optimization method based on the cooperative game of hydrogen storage system, which takes into account the waste heat utilization of the electrolyzer and fuel cell, reduces the total operation cost of the system, and improves the economy of system operation. However, a large number of studies have mainly considered the economic benefits of hydrogen storage system participation in system operation, without considering the thermoelectric–hydrogen coupling characteristics of the fuel cell and the electrolyzer, as well as the lack of comprehensive and refined modeling of the hydrogen energy system and the impact of the wind and light output uncertainty on the system operation.

Renewable energy sources, such as wind and solar, have a high degree of uncertainty and can pose a challenge to the economic operation of integrated energy systems. In the context of uncertainty about optimal scheduling, the prevailing approach is to employ stochastic optimization and robust optimization (RO) methods. The stochastic optimization method typically assumes a probability distribution of uncertain variables. However, due to the inherent complexity of the factors influencing the uncertainty of these variables, the method often fails to accurately reflect the actual pattern. The RO approach is predicated on a worst-case scenario, thereby enhancing the efficiency of the solution. However, this approach entails the loss of a considerable amount of useful data, is characterized by a high degree of conservatism, and results in the sacrifice of economic efficiency. In [13], Bagheri et al. created a two-stage stochastic planning model for the selection and operation of various distributed energy sources in the peaking and valley filling of community buildings. In [14], Zakaria et al. offer a comprehensive overview of the generic steps of stochastic optimization for renewable energy applications, delineating the processes of uncertainty modeling and sampling of relevant information. The authors of references [15–17] have employed the technique of stochastic optimization to arrive at solutions to multi-type energy scheduling problems. In recent years, distributed robust optimization methods have been proposed as a means of addressing the shortcomings inherent to stochastic planning and robust regularization. In [18], Liu et al. put forward an effective and robust game-theoretic methodology aimed at enhancing the storage capacity and minimizing energy consumption in a scenario where distributed generation introduces an element of uncertainty. In [19], Ju et al. proposed a new data-driven two-stage robust optimization scheduling model for virtual power plants (VPPs) considering the equivalent conversion mechanism of carbon-green certificates, which considered the equivalent conversion mechanism of carbon-green certificates. This facilitates the optimal aggregation and utilization of rural distributed energy resources. In [20], Ju et al. devised a collaborative model for emerging energy power stations and shared energy storage to engage in the electricity green certificate market. This involved integrating scenario generation and robust optimization techniques to propose a two-stage RO model with enhanced uncertainty intervals. In [21], Sun et al. established a data-driven distribution robust optimization scheduling model for AC and DC distribution networks, and its validity was verified.

This paper synthesizes the findings of the aforementioned studies and identifies key issues. It considers the combined heat, power, and hydrogen supply characteristics of fuel cells and electrolyzers, develops models for hydrogen to electricity and heat in fuel cells and electricity to hydrogen and heat in electrolyzers, and performs detailed modeling of each component to study the optimal scheduling of combined heat, power, and hydrogen supply type microgrids. Meanwhile, to address the inherent unpredictability of wind power output, a data-driven distributional robust optimization scheduling model for integrated cogeneration-type microgrids, encompassing heat, power, and hydrogen supply, is proposed. This model employs a data-driven distributional robust approach to confront the vagaries of wind power output while incorporating 1-paradigm and ∞-paradigm numbers to constrain the uncertainty probability distribution confidence set simultaneously. Ultimately, the model is validated through an illustrative arithmetic example that underscores the vital necessity to enhance the utilization rate of wind power PV and curtail energy and environmental pollution to stimulate economic growth. A brief summary is given in Table 1.

Table 1. Summarization of the literature review.

Reference	Thermoelectric Hydrogen Coupling	Refined Modeling of Hydrogen Energy Systems	Distributed Robust Optimization
[9]	√	×	×
[10]	√	×	×
[11]	√	×	×
[12]	√	×	×
[21]	×	×	√
[22]	√	√	×
This study	√	√	√

2. An Integrated Energy System Combining Heat, Power, and Hydrogen

2.1. Structure of IES

The configuration of the integrated energy system, which combines heat, power, and hydrogen, is illustrated in Figure 1. The system is primarily constituted by distributed power sources, predominantly based on photovoltaic and wind power generation, energy storage apparatus, energy supply apparatus, and loads. The category of energy storage devices encompasses a variety of systems, including electric energy storage systems, hydrogen energy storage systems, and thermal energy storage systems. These include electrical and thermal loads, among others. The electric boiler can be defined as a thermoelectric coupling element, while the fuel cell and electrolyzer can be defined as thermoelectric hydrogen coupling elements. Interventional studies involving animals or humans, and other studies that require ethical approval, must list the authority that provided approval and the corresponding ethical approval code.

2.2. Mathematical Modeling of a Combined Heat, Power, and Hydrogen-Integrated Energy System

2.2.1. Modeling of Hydrogen Energy Storage Systems

Hydrogen energy is a highly energy-efficient and non-polluting clean energy source that is currently the subject of considerable interest within the scientific community. Nevertheless, the current hydrogen energy system exhibits a low conversion efficiency and a considerable amount of heat loss, which collectively result in the ineffective utilization of energy. Consequently, the utilization of waste heat represents an efficacious method for enhancing the efficiency of hydrogen energy systems. The modeling of the hydrogen energy storage system refers to the literature [22].

The voltage of a proton exchange membrane fuel cell electric stack can be expressed as follows [23]:

$$V\text{cell} = E\text{nerst} - V\text{act} - V\text{ohm} - V\text{con} \tag{1}$$

where $Enerst$ is the thermodynamic electric potential; $Vact$ is the activation polarization overvoltage; $Vohm$ is the ohmic polarization overvoltage; and $Vcom$ is the concentration polarization overvoltage, all in V, which can be expressed as follows [24]:

$$\begin{cases} E_{\text{nerst}} = \frac{\Delta G}{2F} + \frac{T_{\text{fc}} - T_{\text{b}}}{2F}\Delta S + \frac{RT_{\text{fc}}}{2F}\ln(P_{\text{H2}}P_{\text{O2}}^{0.5}) \\ V_{\text{act}} = \xi_1 + \xi_2 T_{\text{fc}} + \xi_4 T_{\text{fc}} \ln I_{\text{fc}} + \\ \xi_3 T_{\text{fc}} \ln \frac{P_{\text{O2}}}{5.08\times 10^6 \exp(-498/T_{\text{fc}})}, \\ V_{\text{ohm}} = I_{\text{fc}} R_{\text{imt}}, \\ V_{\text{con}} = \alpha \exp(\beta I_{\text{fc}}), \end{cases} \tag{2}$$

where ΔG is the Gibbs free energy; F is Faraday's constant; T_{fc} is the fuel cell temperature, which is considered as a constant; T_b is the standard temperature; ΔS is the standard molar entropy; P_{H2} and P_{O2} are the partial pressures of the input hydrogen and oxygen, respectively; R is the gas constant; $\xi_1 \sim \xi_4$ is a constant; I_{fc} is the fuel cell current; R_{imt} is the equivalent impedance of the internal resistance, which is a constant when the temperature is unchanged; and α and β are all constants.

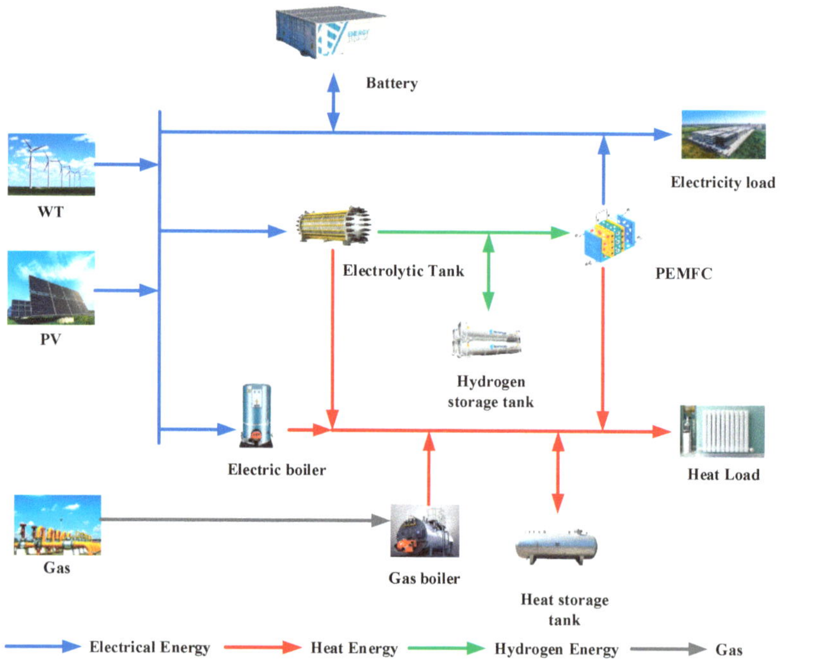

Figure 1. The structural configuration of a combined heat, power, and hydrogen integrated energy system.

The fuel cell efficiency is

$$\eta_{\text{fc}} = \mu \frac{V_{\text{cell}}}{H_{\text{H2}}}, \tag{3}$$

where V_{cell} is the fuel cell voltage; μ is the fuel utilization rate; and H_{H2} is the high calorific value of hydrogen.

The relationship between P_{fc_H2}, P_{fc}, and H_{fc} can be expressed as follows:

$$\begin{cases} P_{\text{fc\_H2}} = P_{\text{fc}} + H_{\text{fc}}, \\ P_{\text{fc}} = \eta_{\text{fc}} P_{\text{fc\_H2}} \end{cases} \tag{4}$$

where P_{fc_H2} represents the total power obtained by the fuel cell from the hydrogen storage tank; P_{fc} is the electrical power; and H_{fc} is the remaining power.

The power transferred from the fuel cell to the DC bus is expressed as follows:

$$\begin{cases} n_{fc} = P_{fc}/(\eta_{fc}L_{H2}) \\ P_{fc\_BUS} = \eta_{fc\_dc}P_{fc} \end{cases} \tag{5}$$

where n_{fc} is the rate of hydrogen consumption by the fuel cell; L_{H2} is the low calorific value of hydrogen; η_{fc_dc} is the efficiency of the fuel cell converter; and P_{fc_BUS} represents the input power of the fuel cell bus.

The relationship between the electrical power output of the fuel cell and the thermal energy can be expressed as follows:

$$H_{fc} = (1 - \eta_{fc})P_{fc}/\eta_{fc} \tag{6}$$

The heat energy generated by the fuel cell is divided into two distinct portions. One is dissipated into the surrounding environment, while the other is transferred to the thermal bus.

$$H_{fc\_BUS} = \eta_{fc\_re}H_{fc} \tag{7}$$

where H_{fc_BUS} represents the thermal power output from the thermal bus and η_{fc_re} is the heat transfer efficiency

The U-I characteristic equation for an alkaline electrolyzer at any temperature is given by the following equation:

$$U_{cell} = V_{rev} + \frac{r_1 + r_2 T_{el}}{A_{cell}} + (s_1 + s_2 T_{el} + s_3 T_{el}^2) \times \lg(\frac{s_4 + s_5/T_{el} + s_6/T_{el}^2}{A_{cell}} I_{el} + 1) \tag{8}$$

where r_1, r_2 is the electrolyte ohmic resistance parameter; T_{el} is the electrolyzer temperature; A_{cell} is the electrolytic module area; I_{el} is the DC; $s_1 \sim s_6$ is the electrode over-voltage coefficient; and V_{rev} is the cell open-circuit voltage.

The electrolyzer efficiency is

$$\eta_{el} = a_1 \exp(\frac{a_2 + a_3 T_{el}}{I_{el}/A_{cell}} + \frac{a_4 + a_5 T_{el}}{(I_{el}/A_{cell})^2}) \tag{9}$$

where $a_1 \sim a_5$ is the Faraday efficiency factor.

$$\begin{cases} P_{el} = P_{el\_H2} + H_{el} \\ P_{el} = P_{el\_H2}/\eta_{el} \end{cases} \tag{10}$$

where P_{el} is the electrical power obtained by the electrolyzer from the DC busbar; P_{el_H2} is the power used for hydrogen production; H_{el} is the power used by the electrolyzer to produce heat; and η_{el} is the efficiency of the electrolyzer.

The rate of hydrogen production in the electrolyzer is as follows:

$$\begin{cases} n_{el} = \eta_{el}P_{el}/H_{HV\_H2} \\ P_{el\_BUS} = P_{el}/\eta_{el\_dc} \end{cases} \tag{11}$$

where n_{el} is the rate of hydrogen production in the electrolyzer; P_{el_BUS} is the electrical power supplied by the DC bus to the electrolyzer; and η_{el_dc} is the efficiency of the electrolyzer converter.

In conclusion, the power transmitted from the electrolyzer to the bus is

$$H_{el\_BUS} = \eta_{el\_re}H_{el} \tag{12}$$

where H_{el_BUS} is the thermal power flowing from the electrolyzer to the thermal busbar and η_{el_re} is the heat transfer efficiency.

Hydrogen tanks are utilized for the storage of hydrogen. According to the van der Waals equation of state for real gases, the pressure within the tank can be expressed as follows:

$$\begin{cases} P_{\text{sto}} = \frac{n_{\text{sto}} R_c K}{V_{\text{sto}} - n_{\text{sto}} b} - \frac{n_{\text{sto}}^2 a}{V_{\text{sto}}^2} \\ n_{\text{sto}} = \int (n_{\text{el}} - n_{\text{fc}}) dt + n_{\text{re}} \\ S_{\text{OHC}} = P_{\text{sto}} / P_{\text{N}} \end{cases} \tag{13}$$

In this context, R_c represents Avogadro's constant (6.02×10^{23}/mol), K denotes the Kelvin temperature, V_{sto} is the volume of the hydrogen storage tank, n_{sto} is the amount of hydrogen in the tank, a and b are the proportionality coefficient, respectively, $a = 0.02476\ \text{Pa} \cdot \text{m}^6/\text{mol}$, $b = 2.661 \times 10^{-5}\ \text{m}^6/\text{mol}$, n_{re} is the initial amount of hydrogen stored in the tank, S_{OHC} is the state of storage of the hydrogen storage tank P_{N} is the maximum pressure, and t represents the moment.

2.2.2. Wind Power Systems

In the process of electricity generation, a wind power system first transforms wind energy into mechanical energy and subsequently converts the mechanical energy into electrical energy [25].

$$P_{\text{W}}(t) = \begin{cases} 0 & v \le v_{in},\ v_{\text{out}} \le v \\ \frac{P_{\text{r}}(v - v_{in})}{v_{\text{r}} - v_{in}} & v_{\text{in}} \le v \le v_{\text{r}} \\ P_{\text{r}} & v_{\text{r}} \le v \le v_{\text{out}} \end{cases} \tag{14}$$

where $P_{\text{W}}(t)$ is the generating power of the turbine; P_{r} is the rated generating power of the turbine; v_{in} is the cut-in wind speed of the turbine; v_{r} is the rated wind speed of the turbine; and v_{out} is the cut-out wind speed of the turbine.

2.2.3. Photovoltaic Power Generation System

The generation of power by a photovoltaic system is influenced not only by its size, capacity, and other intrinsic parameters but also by the actual light intensity and the influence of the external ambient temperature. Consequently, once the parameters have been established, the output power of the photovoltaic cell can be expressed as follows [26]:

$$P_{\text{PV}}(t) = P_{\text{sta}} f_{\text{PV}} \frac{G(t)}{G_{\text{sta}}} \{1 + \alpha_{\text{T}} [T_{\text{a}}(t) - T_{\text{a,sta}}]\} \tag{15}$$

where $P_{\text{PV}}(t)$ is the output power of the PV panel; P_{sta} is the rated power of the PV panel under the standard parameters; f_{PV} is the attenuation coefficient of the power; $G(t)$ is the intensity of solar irradiation; G_{sta} is the intensity of solar irradiation under the standard test conditions; α_{T} is the temperature coefficient of the power of the PV panel; and $T_{\text{a}}(t)$ and $T_{\text{a,sta}}$ are the temperatures of the PV panel and ambient temperatures under the standard test conditions, respectively. The current temperature of the photovoltaic panel can be calculated using the following formula:

$$T_{\text{a}}(t) = T(t) + \lambda G(t) \tag{16}$$

where $T(t)$ represents the ambient temperature and λ denotes the radiant temperature coefficient, which has a value of 0.0256.

2.2.4. Thermal Storage System [27]

$$\begin{cases} H_{\text{tst}} = H_{\text{tst}}(t-1) + \left[H_{\text{tst-ch}}(t) \eta_{\text{tst-ch}} - \frac{H_{\text{tst-dis(t)}}}{\eta_{\text{tst-dis}}} \right] \Delta t \\ H_{\text{oc}}(t) = H_{\text{tst}}(t) / C_{\text{tst}} \times 100\% \end{cases} \tag{17}$$

where H_{tst} is for the thermal storage tank moment t stored thermal energy; $H_{\text{tst-ch}}(t)$ and $H_{\text{tst-dis(t)}}$, respectively, are for the thermal storage tank moment t storage and exothermic power; $\eta_{\text{tst-ch}}$ and $\eta_{\text{tst-dis}}$, respectively, are for the thermal storage tank storage and release thermal efficiency; $H_{oc(t)}$ is for the thermal storage tank at the moment t thermal storage state; and C_{tst} is for the rated capacity of the thermal storage tank.

2.2.5. Electric Energy Storage Systems [27]

An electrical energy storage device can mitigate the impact of voltage fluctuations resulting from the inherent variability of wind power output. This enables the reliable storage, flow, and release of electrical energy. The mathematical model of the battery can be expressed as follows:

$$\begin{cases} E_{\text{bat}}(t) = E_{\text{bat}}(t-1) + \left[P_{\text{bat-ch}}(t)\eta_{\text{bat-ch}} - \frac{P_{\text{bat-dis}}(t)}{\eta_{\text{bat-dis}}}\right]\Delta t \\ S_{\text{OC}}(t) = E_{bat}(t)/C_{\text{bat}} \times 100\% \end{cases} \tag{18}$$

where $E_{\text{bat}}(t)$ is for the battery moment t stored power; $P_{\text{bat-ch}}(t)$ and $P_{\text{bat-dis}}(t)$ are for the battery moment t charging and discharging power, respectively; $\eta_{\text{bat-ch}}$ and $\eta_{\text{bat-dis}}$ are for the battery charging and discharging efficiency, respectively; $S_{\text{OC}}(t)$ is for the battery at the moment t charge state; C_{bat} is for the rated capacity of the battery; and Δt is for the scheduling time interval.

2.2.6. Electric Boiler

An electric boiler is an energy coupling device that converts electrical energy into thermal energy. The efficiency of this conversion is dependent on both the efficiency of the boiler itself and the efficiency of the converter.

$$H_{\text{eb}}(t) = \eta_{\text{eb}} P_{\text{eb}}(t) \tag{19}$$

where $H_{\text{eb}}(t)$ is the heat generated by the electric boiler at the time of t; η_{eb} is the electric boiler electric–thermal conversion efficiency; $P_{\text{eb}}(t)$ is the size of the power absorbed by the electric boiler from the bus at the time of t.

2.2.7. Gas Boiler

$$H_{\text{gb}}(t) = \eta_{\text{gb}} P_{\text{gb}}(t) \tag{20}$$

where $H_{\text{gb}}(t)$ is for the moment t heat produced by the gas boiler; η_{gb} is for the gas boiler heating efficiency; and $P_{\text{gb}}(t)$ is for the moment t gas boiler gas consumption.

3. Optimal Scheduling Model for Daily Operation Based on Hydrogen Energy Storage

3.1. Objective Function

The objective of the day-to-day operation optimization scheduling is to ensure the lowest system operation cost, which includes the costs associated with operation and maintenance, environmental impact, and gas purchase. In light of the aforementioned considerations, the objective function is expressed as follows:

$$\min F = F_{\text{ope}} + F_{\text{buy}} + F_{\text{en}} \tag{21}$$

where F_{ope} is the operation and maintenance cost; F_{buy} refers to the cost of purchased gas.; and F_{en} is for environmental costs.

3.1.1. Operation and Maintenance Costs

$$\min F_{\text{ope}} = \sum_{t=0}^{T_d} \left[\beta_{\text{wt}} P_{\text{W}}(t) + \beta_{\text{pv}} P_{\text{pv}}(t) + \beta_{\text{hss}} P_{\text{hss}}(t) + \beta_{\text{bat}} P_{\text{bat}}\left(t\right) + \beta_{\text{eb}} H_{\text{eb}}(\text{t}) + \beta_{\text{gb}} H_{\text{gb}}(\text{t}) + \beta_{\text{tst}} H_{\text{tst}}(\text{t})\right] \tag{22}$$

where T_d is the scheduling cycle; P_{hss} and P_{bat} are the charging and discharging power of the hydrogen energy storage system and the electric energy storage system at the moment

t, respectively; and β_{wt}, β_{pv}, β_{hss}, β_{bat}, β_{eb}, β_{gb}, and β_{tst} are the unit O&M costs of wind turbines, photovoltaics, hydrogen storage systems, batteries, electric boilers, gas boilers, and thermal storage tanks, respectively.

3.1.2. Cost of Gas Purchase

$$F_{buy} = \sum_{t=0}^{T_d} \frac{R_{gas} H_{gb}(t)}{R_H} \tag{23}$$

where R_{gas} is the unit price of natural gas (CNY/m^3) and R_H is the calorific value of natural gas (kWh/m^3).

3.1.3. Environmental Costs

$$F_{en} = \sum_{t=0}^{T_d} (L_{SO_2}\lambda_{SO_2} + L_{CO_2}\lambda_{CO_2} + L_{NO_2}\lambda_{NO_2}) H_{gb}(t) \tag{24}$$

where L_{SO_2}, L_{CO_2}, L_{NO_2}, and λ_{SO_2}, λ_{CO_2}, λ_{NO_2} are the emission factors and treatment costs of pollutant gases SO_2, CO_2, and NO_2 produced by gas boilers, respectively.

3.2. *Restrictive Condition*

3.2.1. Power Balance Constraints

Electrical power balance constraints and thermal power balance constraints

$$\begin{cases} P_{wt}(t) + P_{pv}(t) + P_{bat\text{-}dis}(t) + P_{fc}(t) = P_{load}(t) + P_{el}(t) + P_{eb}(t) + P_{bat\text{-}ch}(t) \\ H_{load}(t) + H_{tst\text{-}ch}(t) = H_{gb}(t) + H_{eb}(t) + H_{tst\text{-}dis}(t) + H_{el}(t) + H_{fc}(t) \end{cases} \tag{25}$$

where $H_{load}(t)$ is the heat load at time t.

3.2.2. Battery Operation Constraints

$$y^{ch} + y^{dis} \le 1 \tag{26}$$

$$\begin{cases} y^{ch} P_{ch\ min} \le P_{bat\text{-}ch}(t) \le y^{ch} P_{ch\ max} \\ y^{dis} P_{dis\ min} \le P_{bat\text{-}dis}(t) \le y^{dis} P_{dis\ max} \end{cases} \tag{27}$$

$$S_{OC\ min} \le S_{OC}(t) \le S_{OC\ max} \tag{28}$$

where y^{ch} and y^{dis} are for the charging and discharging state, 0–1 variable; $P_{ch\ min}$, $P_{ch\ max}$, $P_{dis\ min}$, and $P_{dis\ max}$ are for the lower and upper limits of charging and discharging power, respectively; and $S_{OC\ min}$, $S_{OC\ max}$ for the lower and upper limits of the battery charging state, respectively.

3.2.3. Thermal Storage Constraints

$$z^{ch} + z^{dis} \le 1 \tag{29}$$

$$\begin{cases} z^{ch} H_{ch\ min} \le H_{tst\text{-}ch}(t) \le z^{ch} H_{ch\ max} \\ z^{dis} H_{dis\ min} \le H_{tst\text{-}dis}(t) \le z^{dis} H_{dis\ max} \end{cases} \tag{30}$$

$$H_{OC\ min} \le H_{OC}(t) \le H_{OC\ max} \tag{31}$$

where z^{ch} and z^{dis}, respectively, are for the state of charge and discharge, 0–1 variables; $H_{ch\ min}$ and $H_{ch\ max}$ as well as $H_{dis\ min}$ and $H_{dis\ max}$, respectively, are for the lower and upper limits of the charge and discharge thermal power; $H_{OC\ min}$ and $H_{OC\ max}$, respectively, are for the lower and upper limits of the thermal storage state of the thermal storage tank.

3.2.4. Restrictions Associated with Hydrogen Energy Systems

$$S_{\mathrm{OHC\ min}} \leq S_{\mathrm{OHC}}(t) \leq S_{\mathrm{OHC\ max}} \tag{32}$$

$$\begin{cases} P_{\mathrm{fc\ min}} \leq P_{\mathrm{fc}}(t) \leq P_{\mathrm{fc\ max}} \\ P_{\mathrm{el\ min}} \leq P_{\mathrm{el}}(t) \leq P_{\mathrm{el\ max}} \\ P_{\mathrm{hy}}(t) = a_{\mathrm{fc}} P_{\mathrm{fc}}(t) + a_{\mathrm{el}} P_{\mathrm{el}}(t) \end{cases} \tag{33}$$

$$a_{\mathrm{fc}} + a_{\mathrm{el}} \leq 1 \tag{34}$$

where $P_{\mathrm{fc\ max}}$ and $P_{\mathrm{fc\ min}}$ are the upper and lower limits of fuel cell power; $P_{\mathrm{el\ max}}$ and $P_{\mathrm{el\ min}}$ are the upper and lower limits of electrolyzer power; $S_{\mathrm{OHC\ max}}$ and $S_{\mathrm{OHC\ min}}$ are the upper and lower limits of hydrogen storage state $S_{\mathrm{OHC}}(t)$; $P_{\mathrm{hy}}(t)$ is the operating work of the hydrogen system at the moment t; and a_{fc} and a_{el} are the working flag bits of the fuel cell and electrolyzer, respectively, which take the value of 0 or 1 whereby 0 is not working and 1 is working.

3.2.5. The Limitations of Power for Each Device

$$\begin{cases} P_{\mathrm{eb\ min}} \leq P_{\mathrm{eb}}(t) \leq P_{\mathrm{eb\ max}} \\ P_{\mathrm{gb\ min}} \leq P_{\mathrm{gb}}(t) \leq P_{\mathrm{gb\ max}} \end{cases} \tag{35}$$

where $P_{\mathrm{eb\ max}}$ and $P_{\mathrm{eb\ min}}$ are the upper and lower limits for electric boilers and $P_{\mathrm{gb\ max}}$ and $P_{\mathrm{gb\ min}}$ are the upper and lower limits for gas boiler power.

3.3. *Data-Driven Distribution Robust Scheduling Models*

By the stipulations outlined in the regulatory framework, the characteristics of the equipment in question, the state of charge and discharge of the battery, the state of charge and discharge of the heat storage tank, the state of charge and discharge of the hydrogen storage tank, the operational status of the electric boiler, the operational status of the gas boiler, the casting and cutting of the electrolyzer, and the casting and cutting of the fuel cell are collectively designated as the initial stage variables, x. The remaining continuous decision variables are of the second-stage type, designated as y_s. The deterministic model (1)–(35) can be expressed in the following uncertainty form when employing multiple scenarios to describe scenery uncertainty.

$$\min_{x, y_s \in Y} \sum_{s=1}^{N_s} p_s A^T y_s \tag{36}$$

$$\text{s.t.} \begin{cases} Cx \leq c \\ Dx = d \end{cases} \tag{37}$$

$$\begin{cases} Ey_s \leq e \\ Fy_s = f \end{cases} \tag{38}$$

$$\begin{cases} Gx + Hy_s \leq g \\ Kx + My_s = u \end{cases} \tag{39}$$

where p_s is the correlation probability under s scenarios; N_s is the total number of scenarios; Equation (32) denotes the correlation constraints of the first-stage variables, which specifically correspond to Equations (27), (30), (35) and (36); and Equation (39) represents the second-stage variable-related constraints, which specifically correspond to the capacity constraints for the batteries, thermal storage tanks, and hydrogen storage tanks as expressed in Equations (28), (31), and (32), as well as the output constraints for the electric boiler and gas boiler, as expressed in Equation (36). Equation (40) represents the coupled constraint relationship between the first and second stage variables, which specifically corresponds to the power constraints for battery and heat storage tank charging and discharging as expressed in Equations (27) and (30), as well as the output constraints for electric and gas boilers as expressed in Equation (35).

Assuming that there are M actual scenarios, M_1, M_2..., M_{Ns} finite discrete scenarios satisfying $M = \sum_1^{N_s} M_{Ns}$ are obtained by clustering screening, a theoretical distribution of uncertainty variables $p_0 = (p_0^1, p_0^2, \ldots p_0^{Ns})$ can be constructed, where $p_0^{Ns} = M_{Ns}/M$. However, as the initial distribution obtained through historical data differs from the actual distribution, it is essential to establish confidence sets for varying values of θ, as illustrated in Equation (40), as follows:

$$\Omega^{\mathrm{P}} = \left\{ \{p_s\} \left| \begin{array}{l} p_s \geq 0, s = 1, \ldots, N_s \\ \sum\limits_{s=1}^{N_s} p_s = 1 \\ \|p - p_0\|_1 \leq \theta_1 \\ \|p - p_0\|_{\mathbb{R}\infty} \leq \theta_\infty \end{array} \right. \right\} \tag{40}$$

where θ_1 and θ_∞ represent the acceptable limits of probability under the specified constraints of the 1-paradigm and ∞-paradigm, respectively. Ω^{P} denotes the p_s-feasible domain, whereas p denotes a vector with element p_s. The construction of the confidence set is dependent upon two key factors: the specified confidence level and the amount of historical data. This is illustrated by Equation (41).

$$\left\{ \begin{array}{l} \Pr\{\|p - p_0\|_1 \leq \theta_1\} \geq 1 - 2N_s \mathrm{e}^{\frac{-2M\theta_1}{N_s}} \\ \Pr\{\|p - p_{\mathbb{Z}\infty}\|_1 \leq \theta_\infty\} \geq 1 - 2N_s \mathrm{e}^{-2M\theta_\infty} \end{array} \right. \tag{41}$$

Equating the right-hand side of the inequality in Equation (41) to α_1 and α_∞, respectively, yields Equation (42).

$$\left\{ \begin{array}{l} \theta_1 = \frac{N_s}{2M} \ln \frac{2N_s}{1-\alpha_1} \\ \theta_\infty = \frac{1}{2M} \ln \frac{2N_s}{1-\alpha_\infty} \end{array} \right. \tag{42}$$

As illustrated in Equations (40)–(42), an increase in historical data results in a reduction in the value of θ. This implies that the confidence set, maintained at the same confidence level, contracts with the accumulation of historical data. Upon the addition of ∞ units of historical data, the values of θ_1 and θ_∞ reach zero, suggesting that the initial distribution accurately represents the true distribution.

The wind and light uncertainties are addressed through the application of the methodologies delineated in Equations (40)–(42), thereby yielding a data-driven distributionally robust optimized scheduling-based model as illustrated in Equation (43).

$$\left\{ \begin{array}{l} \min\limits_{x \in X} \max\limits_{p_s \in \Omega^P} \sum\limits_{s=1}^{N_s} p_s \min\limits_{y_s \in Y} A^{\mathrm{T}} y_s \\ \text{s.t.Equations (37)–(39)} \end{array} \right. \tag{43}$$

3.4. Distributed Robust Model Solving

To solve Equation (43), the model is decomposed into a master problem (MP) and a sub-problem (SP). This is conducted to use the column and constraint generation algorithm iteratively. The master problem is shown in Equation (44). The optimal solution that satisfies the constraints is solved with a known probability distribution. The lower bound value is given to Equation (43), as follows:

$$\text{MP:} \quad \min_{x \in X, y_s^m \in Y, W} \quad W \tag{44}$$

$$W \geq \sum_{s=1}^{N_s} p_s^m A^{\mathrm{T}} y_s^m, \forall m = 1, 2, \ldots, n \tag{45}$$

In this context, the variable "m" represents the number of iterations.

The subproblem identifies the worst-case probability distribution under real-time operation, given the initial variable x^* in the primary problem, and returns it to the primary problem. Additionally, it provides an upper bound for Equation (43):

$$\mathrm{SP}: f(x^*) = \max_{p_s \in \Omega^{\mathrm{P}}} \sum_{s=1}^{N_s} p_s \min_{y_s \in Y} A^{\mathrm{T}} y_s \tag{46}$$

The inner second-stage variables in Equation (46) are independent of the outer discrete scenario probability values. Consequently, the inner minimization problem can be solved before solving the outer problem.

The particular solution procedure is as follows:

(1) The lower bound should be set to $L_{\mathrm{B}} = 0$, the upper bound to $U_{\mathrm{B}} = \infty$, the number of iterations to 1, and the initial probability distribution applied;

(2) The objective is to solve MP, thereby obtaining the optimal solution (x^*, W^*) and subsequently updating the lower bound $\max\{L_{\mathrm{B}}, W^*\}$;

(3) Maintaining the first stage variable x^* at a constant value, one must solve the subproblem to obtain the worst-case scenario probability distribution p_s^* and the objective function value $f(x^*)$. Furthermore, the upper bound value $\min\{U_{\mathrm{B}}, f(x^*)\}$ must be updated by the aforementioned results;

(4) If $U_{\mathrm{B}} - L_{\mathrm{B}} < \varepsilon$, the iteration should be terminated and the optimal solution x^* should be reinstated. Conversely, p_s^{m+1} should be updated to equal p_s^*, and the aforementioned steps should be repeated.

4. Results and Discussion

The study focuses on a typical summer day in a park located in northern China. The data samples were generated at random using a normal distribution based on the predicted values. The probability distribution of the initial scenes is obtained through a probability distance-based scene reduction technique, resulting in a total of four scenes. The IES internal electrical and thermal load curves, as well as the wind and PV output data for Scenarios 1 through 4, are shown in Figures 2–5. The internal equipment parameters of the integrated energy system are listed in Table 2.

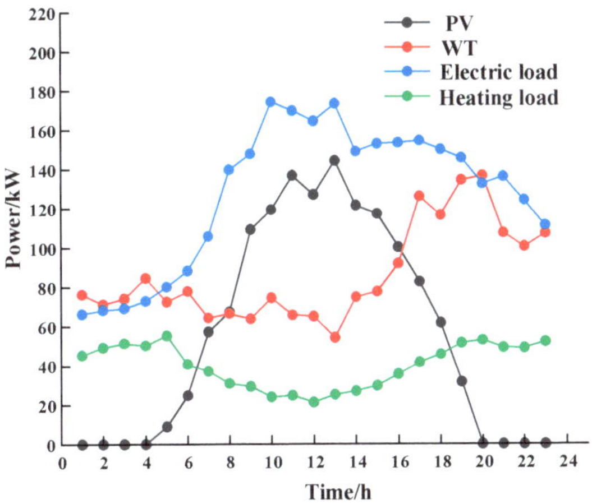

Figure 2. Scenario 1 Electrical and Thermal Loads and their Wind and PV Power Curves.

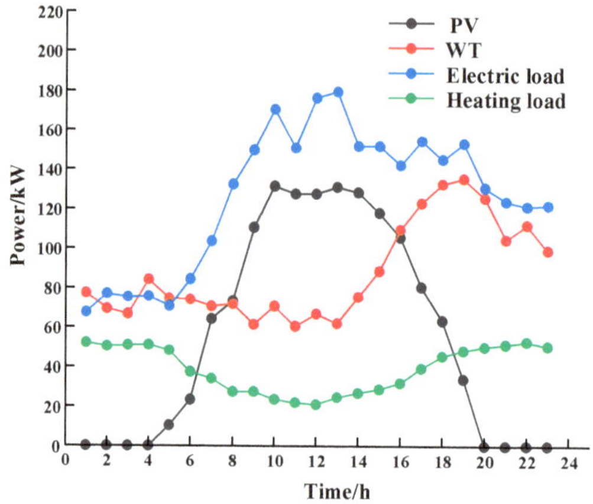

Figure 3. Scenario 2 Electrical and Thermal Loads and their Wind and PV Power Curves.

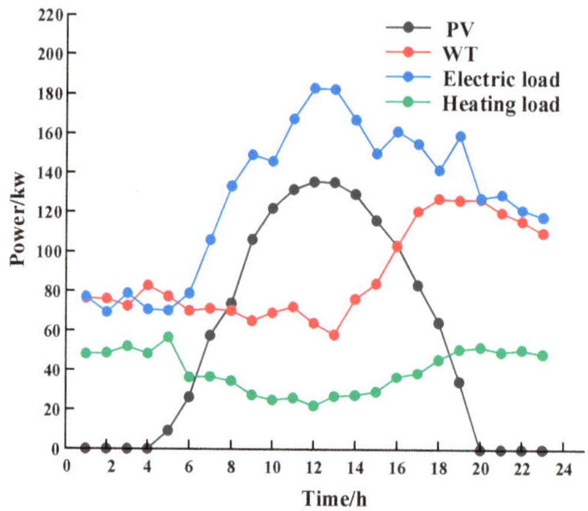

Figure 4. Scenario 3 Electrical and Thermal Loads and their Wind and PV Power Curves.

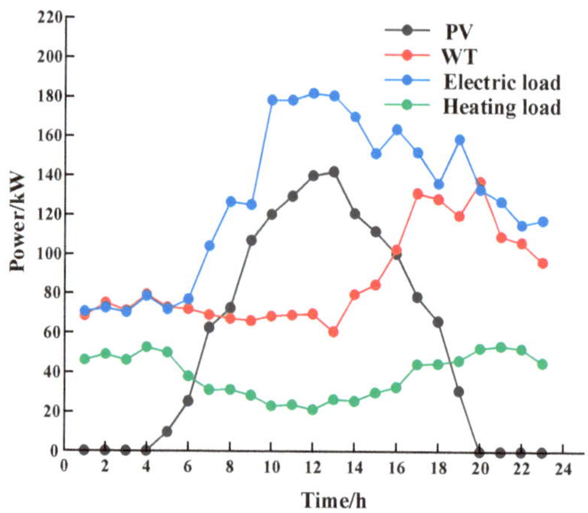

Figure 5. Scenario 4 Electrical and Thermal Loads and their Wind and PV Power Curves.

Table 2. Value of the parameters.

Parament	Value	Parament	Value
η_{eb}	0.9	$P_{ch\ min}$, $P_{ch\ max}$	0, 100 kw
η_{gb}	0.73	$P_{dis\ min}$, $P_{dis\ max}$	0, 100 kw
$\eta_{tst\text{-}ch}$, $\eta_{tst\text{-}dis}$	0.9	$H_{ch\ min}$, $H_{ch\ max}$	0, 60 kw
$\eta_{bat\text{-}ch}$, $\eta_{bat\text{-}dis}$	0.9	$H_{dis\ min}$, $H_{dis\ max}$	0, 60 kw
η_{fc}	0.45	$P_{fc\ min}$, $P_{fc\ max}$	0, 50 kw
η_{el}	0.55	$P_{el\ min}$, $P_{el\ max}$	0, 60 kw
P_N	30 Mpa	C_{bat}	300 kwh
V_{sto}	28 L	C_{tst}	300 kwh
$P_{eb\ min}$, $P_{eb\ max}$	0, 60 kw	$P_{gb\ min}$, $P_{gb\ max}$	0, 70 kw

4.1. Supply and Demand Balance Analysis

The power–heat balance diagram for optimal scheduling of a cogeneration-type microgrid containing hydrogen storage based on distributed robust optimization is analyzed in Scenario 1. The remaining scenarios are similar to this one and will not be repeated here for the sake of brevity. The system employs both a wind power system and a photovoltaic power system as its primary source of energy. As illustrated in Figure 6a, during the daytime, the surplus electricity is utilized for electrolyzing water, thereby producing hydrogen, and a modest quantity of electricity is stored in the battery. The excess power generated between the hours of 9:00 and 18:00 is utilized for the electrolysis of water, resulting in the production of hydrogen. This is complemented by fuel cell power generation from 20:00 to 23:00 and 1:00 to 5:00 during the nighttime hours when the wind power system is unable to meet the required energy demand. The process of heat balancing is analogous to that of electrical balancing. As illustrated in Figure 6b, during the daytime, the heat required by the system is primarily supplied by the gas boiler, with any excess heat being absorbed by the thermal energy storage system for release at night. The heat load demand at night is relatively high in comparison to the daytime and is met by a variety of heating sources, including gas boilers, electric boilers, fuel cells, electrolysis tanks, and heat storage tanks. The gas boilers are the primary source of heat.

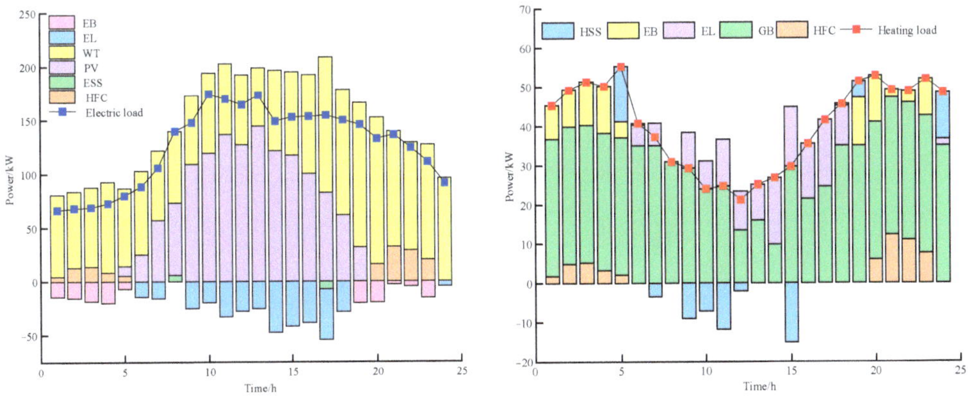

(a) (b)

Figure 6. *Cont.*

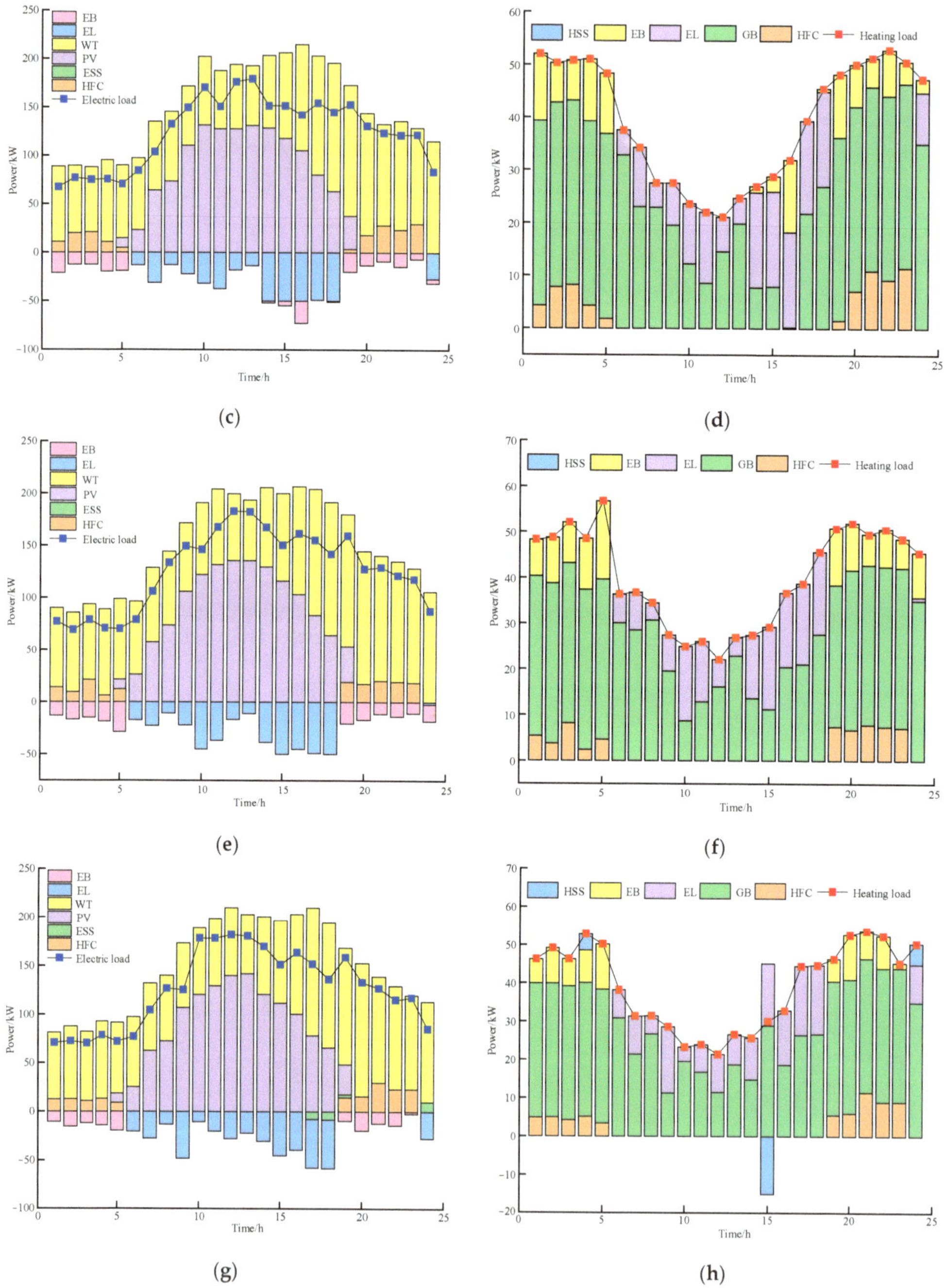

Figure 6. Results of electricity, heat balance scheduling: (**a**) Electrical Balance Diagram for Scene 1; (**b**) Heat Balance Diagram for Scene 1; (**c**) Electrical Balance Diagram for Scene 2; (**d**) Heat Balance Diagram for Scene 2; (**e**) Electrical Balance Diagram for Scene 3; (**f**) Heat Balance Diagram for Scene 3; (**g**) Electrical Balance Diagram for Scene 4; (**h**) Heat Balance Diagram for Scene 4.

4.2. Optimization Results for the DRO and Related Comparative Analysis

4.2.1. Analysis of the Impact of Different Numbers of Historical Data on the Optimization Results

In this section, the impact of varying the number of historical data points on the optimization results in Equation (42) is investigated. For each data point, the confidence levels α_1 and α_∞ are set to 0.9. Based on the prediction scenarios, it is assumed that the prediction errors of WT, PV, and load follow a normal distribution with a mean of the predicted output and a variance of 0.05 times the predicted output. Subsequently, 100–800 scenarios are randomly generated and clustered to obtain four typical scenarios, which are then analyzed. The results obtained are presented in Table 3. As the number of historical scenarios increases, the running cost gradually decreases. This is because as the number of historical scenarios increases, the information on the initial probability distribution is closer to the real value, and the feasible domain of the probability distribution value becomes smaller. This reduces the conservatism of the solution and improves the economic efficiency.

Table 3. Comparison of running cost results for different numbers of historical data.

Data Element	Running Cost /CNY
100	472.4200
200	471.5667
500	471.0546
800	470.9266

4.2.2. An Examination of the Influence of Varying Degrees of Confidence on the Outcomes of Optimization Processes

Similarly, as illustrated in Equation (42), the optimization outcomes vary when the confidence levels α_1 and α_∞ are disparate. The comparative outcomes are presented in Table 4, wherein the historical data set comprises 500 instances. As evidenced in Table 4, an increase in both α_1 and α_∞ results in a corresponding rise in the cost of running the system. This is attributable to several factors, including an elevated confidence level, a larger confidence interval, and a greater degree of uncertainty associated with the random variables included in the model. Consequently, the economic expectations associated with the system become more pronounced.

Table 4. Comparison of running cost results at different confidence sets.

α_1	Running Cost /CNY		
	$\alpha_\infty = 0.5$	$\alpha_\infty = 0.9$	$\alpha_\infty = 0.99$
0.2	472.7815	473.7219	473.8523
0.5	472.8730	473.8134	474.4930
0.99	472.8730	474.1267	475.9204

4.2.3. Comparative Analysis with Deterministic Methods

This section presents a comparative analysis of the decision-making results obtained from the distribution robust optimization model and the traditional deterministic model. The optimization results obtained by the two methods are compared by generating 500 scenarios through random simulation, and the results are presented in Table 5. In the DRO method, both α_1 and α_∞ are set to 0.8.

Table 5. Comparison of the running cost between the DRO and deterministic mode.

	Parameters and Units	Results Under Different Models DRO	Deterministic Model
Miscellaneous costs/CNY	Operation and maintenance costs /CNY	69.2737	69.8616
	Fuel costs /CNY	346.3680	373.2735
	Pollutant gas treatment costs /CNY	55.3590	59.6592
	Running Cost /CNY	471.0007	502.7943

As evidenced in Table 5, the implementation of the DRO methodology results in reduced fuel and pollutant gas treatment costs in comparison to the deterministic approach. Moreover, the operating costs associated with the DRO method are less than those of the deterministic model. This is due to the fact that, in consideration of the inherent uncertainty and fluctuations associated with wind and photovoltaic power, the system is designed with a more conservative approach, whereby a corresponding margin is incorporated into the system. Concurrently, the incorporation of hydrogen storage systems, electric storage, thermal storage, and electric and gas boilers can enhance the alignment of electric and thermal loads and energy output, while reducing operational expenses. Furthermore, as the DRO method also accounts for the double uncertainty inherent in the probability distribution of the scenario, the lowest probability distribution expectation is obtained, thereby exhibiting optimal economic performance.

5. Discussion

In this study, through detailed modeling of the hydrogen energy storage system that considers the thermoelectric coupling characteristics of fuel cells and electrolyzers, the system achieves economically stable operation and reduces energy waste. The established comprehensive energy system with hydrogen storage can operate independently to meet the electrical and thermal load demands of the system. Section 4.2.1 of the study examines the impact of different numbers of historical data on operating costs. It is observed that as the number of historical scenarios increases, the operating costs gradually decrease. This is because with more historical scenarios, the initial probability distribution information becomes closer to the true value, narrowing the feasible domain of the probability distribution value, reducing the conservatism of the solution, and resulting in better economic performance. Section 4.2.2 investigates the impact of different confidence levels on the optimization results. As the confidence interval increases, the uncertainty of the random variables in the model increases, leading to a worse economic expectation and higher operating costs. The final part of the study compares deterministic optimization with distributed robust optimization, demonstrating that distributed robust optimization exhibits the best economic performance, reducing operating costs by 6% compared to deterministic optimization and achieving a balance between economic efficiency and robustness.

6. Conclusions

This paper presents the construction of an energy system based on the energy storage model, comprising an electric energy system, a hydrogen energy system, and a thermal energy system. Additionally, it proposes an optimized scheduling model of a combined heat and power type integrated energy system based on distributed robust optimization of hydrogen energy storage. Through the analysis of arithmetic examples, the following conclusions are derived:

1. The integrated energy system constructed in this paper is designed to leverage the complementary advantages and coordinated operation between electric, thermal, and hydrogen energy sources, thereby ensuring the satisfaction of electric and thermal load demands within the microgrid and enabling the system to function as an independent power source;

2. The incorporation of fuel cells and electrolyzers into the system scheduling process as cogeneration devices enhances the overall efficiency of the hydrogen energy system. This is achieved by ensuring a consistent and reliable supply of energy to the electric and thermal loads, while simultaneously reducing energy wastage and improving the system's energy utilization rate;
3. The distributed robust optimization method fully integrates the economic benefits of stochastic planning with the conservative attributes of robust optimization. This approach demonstrates superior economic performance compared to the deterministic method, achieving a balance between economic efficiency and robustness.

In our future research, we plan to explore the impact of uncertainty factors on system performance to determine how to enhance its efficiency. We believe it is essential to conduct detailed modeling of the system to better understand and predict the effects of these uncertainties. Moreover, integrating multi-timescale optimization scheduling can help illuminate optimization strategies across different time scales, which is a promising area for further investigation. We also advocate for a more comprehensive analysis of the system's capacity configuration, as this can significantly improve the practicality and application effectiveness of integrated energy systems. After thorough research, we anticipate that this approach will not only enhance the system's economic efficiency but also strengthen its ability to handle uncertain conditions. Therefore, our future work will focus on these aspects to provide more practical and valuable solutions for optimizing the scheduling of integrated energy systems.

Author Contributions: Methodology, J.L. and A.B.; Formal analysis, Y.K.; Investigation, W.L.; Data curation, Y.W.; Writing—original draft, L.X.; Writing—review and editing, Z.Y. and L.X.; Supervision, Z.Y. and W.L.; Project administration, Y.W. All authors have read and agreed to the published version of the manuscript.

Funding: This research was funded by the science and technology program of State Grid Xinjiang Power Co., Ltd. (B330DK240008).

Data Availability Statement: The original contributions presented in the study are included in the article; further inquiries can be directed to the corresponding authors.

Conflicts of Interest: Authors Lei Xu, Yang Kou, Weile Liang, Alihan Bieerke, and Yunshan Wang were employed by the Electric Power Research Institute of State Grid Xinjiang Electric Power Co., Ltd. The remaining author declares that the research was conducted in the absence of any commercial or financial relationships that could be construed as a potential conflict of interest. Electric Power Research Institute of State Grid Xinjiang Electric Power Co., Ltd. had no role in the design of the study; in the collection, analyses, or interpretation of data; in the writing of the manuscript, or in the decision to publish the results.

References

1. Qin, M.; Yang, Y.; Zhao, X.; Xu, Q.; Yuan, L. Low-carbon economic multi-objective dispatch of integrated energy system considering the price fluctuation of natural gas and carbon emission accounting. *Prot. Control Mod. Power Syst.* **2023**, *8*, 61. [CrossRef]
2. Qiu, Y.; Li, Q.; Ai, Y.; Chen, W.; Benouzid, M.; Liu, S. Two-stage distributionally robust optimization-based coordinated scheduling of integrated energy system with electricity-hydrogen hybrid energy storage. *Prot Control. Mod. Power Syst.* **2023**, *8*, 33. [CrossRef]
3. Yang, H.; Zhang, X.; Ma, Y.; Zhang, D. Critical peak rebate strategy and application to demand response. *Prot. Control Mod. Power Syst.* **2021**, *6*, 28. [CrossRef]
4. Lu, Q.; Liu, M. A multi-criteria compromise ranking decision-making approach for analysis and evaluation of community-integrated energy service system. *Energy* **2024**, *306*, 132439. [CrossRef]
5. Dong, Y.; Han, Z.; Li, C.; Ma, S.; Ma, Z. Research on the optimal planning method of hydrogen-storage units in wind–hydrogen energy system considering hydrogen energy source. *Energy Rep.* **2023**, *9*, 1258–1264. [CrossRef]
6. Fang, X.; Dong, W.; Wang, Y.; Yang, Q. Multiple time-scale energy management strategy for a hydrogen-based multi-energy microgrid. *Appl. Energy* **2022**, *328*, 120195. [CrossRef]
7. Song, Y.; Mu, H.; Li, N.; Wang, H.; Kong, X. Optimal scheduling of zero-carbon integrated energy system considering long-and short-term energy storages, demand response, and uncertainty. *J. Clean. Prod.* **2023**, *435*, 140393. [CrossRef]

8. Dong, H.; Shan, Z.; Zhou, J.; Xu, C.; Chen, W. Refined modeling and co-optimization of electric-hydrogen-thermal-gas integrated energy system with hybrid energy storage. *Appl. Energy* **2023**, *351*, 121834. [CrossRef]
9. Liu, N.; Zhang, K.; Zhang, K. Coordinated configuration of hybrid energy storage for electricity-hydrogen integrated energy system. *J. Energy Storage* **2024**, *95*, 112590. [CrossRef]
10. Wang, Y.; Liu, C.; Qin, Y.; Wang, Y.; Dong, H.; Ma, Z.; Lin, Y. Synergistic planning of an integrated energy system containing hydrogen storage with the coupled use of electric-thermal energy. *Int. J. Hydrogen Energy* **2023**, *48*, 15154–15178. [CrossRef]
11. Shi, M.; Wang, H.; Lyu, C.; Xie, P.; Xu, Z.; Jia, Y. A hybrid model of energy scheduling for integrated multi-energy microgrid with hydrogen and heat storage system. *Energy Rep.* **2021**, *7*, 357–368. [CrossRef]
12. Wang, Y.; Qin, Y.; Ma, Z.; Wang, Y.; Li, Y. Operation optimisation of integrated energy systems based on cooperative game with hydrogen energy storage systems. *Int. J. Hydrogen Energy* **2023**, *48*, 37335–37354. [CrossRef]
13. Bagheri, F.; Dagdougui, H.; Gendreau, M. Stochastic optimization and scenario generation for peak load shaving in Smart District microgrid: Sizing and operation. *Energy Build.* **2022**, *275*, 112426. [CrossRef]
14. Zakaria, A.; Ismail, F.B.; Lipu, M.H.; Hannan, M.A. Uncertainty models for stochastic optimization in renewable energy applications. *Renew. Energy* **2020**, *145*, 1543–1571. [CrossRef]
15. Abunima, H.; Park, W.H.; Glick, M.B.; Kim, Y.S. Two-stage stochastic optimization for operating a renewable-based microgrid. *Appl. Energy* **2022**, *325*, 119848. [CrossRef]
16. Yan, Y.; Huang, C.; Guan, J.; Zhang, Q.; Cai, Y.; Wang, W. Stochastic optimization of solar-based distributed energy system: An error-based scenario with a day-ahead and real-time dynamic scheduling approach. *Appl. Energy* **2024**, *363*, 123099. [CrossRef]
17. Li, Y.; Zhang, J.; Wu, X.; Shen, J.; Maréchal, F. Stochastic-robust planning optimization method based on tracking-economy extreme scenario tradeoff for CCHP multi-energy system. *Energy* **2023**, *283*, 129025. [CrossRef]
18. Liu, X.; Ji, Z.; Sun, W.; He, Q. Robust game-theoretic optimization for energy management in community-based energy system. *Electr. Power Syst. Res.* **2023**, *214*, 108939. [CrossRef]
19. Ju, L.; Lv, S.; Zhang, Z.; Li, G.; Gan, W.; Fang, J. Data-driven two-stage robust optimization dispatching model and benefit allocation strategy for a novel virtual power plant considering carbon-green certificate equivalence conversion mechanism. *Appl. Energy* **2024**, *362*, 122974. [CrossRef]
20. Ju, L.; Bai, X.; Li, G.; Gan, W.; Qi, X.; Ye, F. Two-stage robust transaction optimization model and benefit allocation strategy for new energy power stations with shared energy storage considering green certificate and virtual energy storage mode. *Appl. Energy* **2024**, *362*, 122996. [CrossRef]
21. Sun, X.; Qiu, X.; Zhang, Z.; Ren, H.; Zhang, M. Distributed Robust Optimal Dispatching of AC/DC Distribution Network Based on Data Driven Mode. *Power Syst. Technol.* **2021**, *45*, 4768–4778.
22. Li, Q.; Zou, X.; Pu, Y.; Chen, W.; Zhao, S. Optimal Schedule of Combined Heat-Power Microgrid Based on Hydrogen Energy Storage. *J. Southwest Jiaotong Univ.* **2021**, *58*, 9–21.
23. Hu, Q.; Lin, J.; Zeng, Q.; Fu, C.; Li, J. Optimal control of a hydrogen microgrid based on an experiment validated P2HH model. *IET Renew. Power Gener.* **2020**, *14*, 364–371. [CrossRef]
24. Zou, X. Optimization and Scheduling for Electric-Heat-Hydrogen Integrated Energy System of Industrial Parks. Master's Thesis, Southwest Jiaotong University, Chengdu, China, 2022.
25. Hu, K.; Wang, B.; Cao, S.; Li, W.; Wang, L. A novel model predictive control strategy for multi-time scale optimal scheduling of integrated energy system. *Energy Rep.* **2022**, *8*, 7420–7433. [CrossRef]
26. Zhang, J.; Kan, S.; Hua, Y.; Zhang, X.; You, G.; Zhang, Z. Capacity Optimization of CHP Microgrid Based on Hydrogen Energy Storage. *Acta Energiae Solaris Sin.* **2022**, *43*, 428.
27. Cao, Y.; Wang, Q.; Du, J.; Nojavan, S.; Jermsittiparsert, K.; Ghadimi, N. Optimal operation of CCHP and renewable generation-based energy hub considering environmental perspective: An epsilon constraint and fuzzy methods. *Sustain. Energy Grids Netw.* **2019**, *20*, 100274. [CrossRef]

electronics

MDPI

Article

Multi-Objective Optimal Configuration of Hydrogen Fuel Cell-Based Multi-Energy Microgrid System Considering Exergy

Ji Li [1,*], Lei Xu [1], Yang Kou [1], Weile Liang [1], Yunshan Wang [1] and Zhi Yuan [2,*]

1 Electric Power Research Institute of State Grid Xinjiang Electric Power Co., Ltd., Urumqi 830011, China
2 Engineering Research Center of Renewable Energy Power Generation and Grid-Connected Control, Ministry of Education, Xinjiang University, Urumqi 830017, China
* Correspondence: lj24sg@163.com (J.L.); yz2019@xju.edu.cn (Z.Y.)

Abstract: Relying solely on electrical energy storage for energy regulation makes it difficult to provide a stable and efficient energy supply for microgrid systems currently. Additionally, the economic cost of microgrids and the rate of energy use present a challenge that must be addressed. A strategy for allocating capacity for multi-energy microgrids that takes energy efficiency and hydrogen energy into account is offered as a solution to the aforementioned issues. Initially, the construction of the multi-energy microgrid system takes into account the thermoelectric coupling properties of hydrogen energy devices. Second, the system's energy utilization level is measured using the exergy efficiency analysis. Next, the multi-objective capacity optimization allocation model of the multi-energy microgrid system is established, with the exergy efficiency and system economic cost serving as the objective functions. Lastly, the multi-objective model is solved using the ε-constraint approach to find the Pareto frontier, and Technique for Order Preference by Similarity to an Ideal Solution is employed for decision-making. The example results demonstrate that, when compared to a traditional microgrid using electric energy storage, the proposed model can effectively lower the system's economic cost and improve exergy efficiency. Additionally, multi-objective capacity optimization can be used to strike a balance between exergy efficiency and the system's economic cost. For relevant studies on the capacity allocation of multi-energy microgrids, this work can be a helpful resource.

Keywords: multi-energy microgrid; hydrogen; multi-objective optimization; exergy

Citation: Li, J.; Xu, L.; Kou, Y.; Liang, W.; Wang, Y.; Yuan, Z. Multi-Objective Optimal Configuration of Hydrogen Fuel Cell-Based Multi-Energy Microgrid System Considering Exergy. *Electronics* **2024**, *13*, 3800. https://doi.org/10.3390/electronics13193800

Academic Editor: Carlos Andrés García-Vázquez

Received: 8 September 2024
Revised: 22 September 2024
Accepted: 24 September 2024
Published: 25 September 2024

1. Introduction

Traditional energy systems are under pressure to change and improve due to the world's increasing energy demand and environmental issues. Combining hydrogen energy, an efficient and clean secondary energy source, with a multi-energy microgrid system can increase energy utilization efficiency as well as light and wind dissipation capabilities, which is crucial for achieving sustainable energy development [1,2].

Microgrids that combine many energy sources, conversion devices, and loads based on the energy source's properties and the load demand are known as multi-energy microgrids. The key to ensuring the steady and cost-effective operation of multi-energy microgrids is building a sensible configuration and scheduling plan. In [3], Liu et al. studied a multi-energy microgrid which examined the capacity allocation and combined advantages of hybrid electric–thermal shared energy storage inside a microgrid system from both economic and environmental viewpoints. In [4], Shen et al. built a multi-energy microgrid that is connected to gas, electricity, and heat, and suggested an ideal allocation strategy that considered the multi-energy coupled demand response. This enhances the system's economy and environmental friendliness. In [5], Lu et al. built a standalone multi-energy microgrid using electric storage, photovoltaic cells, an electric chiller, an electric boiler, and a wind turbine. They also suggested a two-stage hybrid decision-making approach to determine how best to configure the microgrid. In [6], Chen et al. developed a coordinated scheduling and optimized scheduling model for a multi-energy microgrid that is based on cogeneration and P2G technology. This model increases the multi-energy microgrid's flexibility to satisfy a variety of energy demands. In [7], Chen et al. suggest the optimal scheduling model for a multi-energy microgrid system based on carbon capture and storage technologies. This effectively increases the multi-energy microgrid system's environmental friendliness and efficiency of energy use. The aforementioned reference delves further into the optimal configuration and dispatch of multi-energy microgrid systems to attain the synergistic interplay among diverse energy sources inside the system. Nevertheless, none of them gave thought to integrating hydrogen energy into a multi-energy microgrid. Combining hydrogen energy storage with multi-energy microgrids will benefit the system's economic cost and energy supply stability because of its lengthy storage cycles and large storage capacity [8–10].

A microgrid system with hydrogen–electric hybrid energy storage was created by Li et al. in [11], and the capacity configuration of the system's energy storage components was tuned to optimize the economy and dependability of the system. A two-tier optimal scheduling model for hydrogen-based microgrid systems is proposed by Nguyen et al. [12], which effectively lowers the carbon emissions and dispatch costs of microgrid systems. A microgrid system with hydrogen storage that is akin to cogeneration was created by Nojavan et al. in [13]. Simultaneously, an optimal dispatch strategy was suggested to address the uncertainty of renewable energy sources and loads, thereby improving the stability and flexibility of the microgrid system's energy supply. The aforementioned study did not take waste heat from the hydrogen fuel cell (HFC) and electrolyzer (EL) into account, even though it integrated hydrogen energy into a microgrid system. Currently, hydrogen energy systems have a low conversion efficiency. Because some energy is lost as heat in EL and HFC throughout the electricity–hydrogen–electricity conversion process, using waste heat becomes an effective technique to boost the conversion efficiency of hydrogen energy systems. By using an HFC-based cogeneration unit, the hydrogen-based multi-energy microgrid designed by Fang et al. [14] increases the microgrid system's energy utilization efficiency. In [15], Shi et al. take into account the thermal energy consumption of hydrogen energy devices and offer a multi-energy microgrid scheduling model that combines thermal energy and hydrogen energy storage. Nevertheless, the aforementioned research fails to maximize the microgrid system's energy usage efficiency, making it challenging to coordinate the system's economic costs and energy use.

The majority of current research on the energy utilization analysis of multi-energy flow systems focuses on energy efficiency analysis based on the first law of thermodynamics [16–19]. However, in multi-energy flow systems, different types of energy sources are coupled together. The energy efficiency analysis based on the first law of thermodynamics does not account for the varying quality of these energy sources, making it challenging to accurately assess each type of energy flow [20]. Improving energy quality is a more significant goal than preserving energy quantity when it comes to multi-energy flow system energy efficiency. The same quantity of energy might have distinct qualities since this quality differs among different sources of energy. Based on the second law of thermodynamics, exergy efficiency makes assessing the quality of energy easier by examining the most work that can be conducted with a given amount of energy. This approach considers both the "quality" and the "quantity" of the energy [21]. Exergy efficiencies have been used in research to assess and analyze multi-energy flow systems [22,23], but fewer studies have looked at exergy efficiencies in relation to the best way to allocate capacity in multi-energy microgrids.

In this paper, in order to solve the problems mentioned above and realize the efficient and stable energy supply of the multi-energy microgrid system with hydrogen storage, as well as the balance between the economic cost and energy utilization, firstly, a multi-energy microgrid system including a wind turbine, photovoltaic cells, hydrogen storage, a heat storage tank, an absorption chiller, and an electric chiller is established by considering the application of EL and HFC waste heat. Second, taking into account the system's economic cost as well as the relationship between various energy sources, the multi-objective capacity optimization allocation model of a multi-energy microgrid system is established. The objective function is the system's economic cost and exergy efficiency. Once more, the Pareto frontier of the multi-objective optimization model is solved using the ε-constraint method, and the Technique for Order Preference by Similarity to an Ideal Solution (TOPSIS) is used for decision-making. Ultimately, a case simulation is used to confirm the efficacy of the optimization technique presented in this research.

This paper's Section 2 gives the mathematical model of the multi-energy microgrid system. A multi-objective model of the system's capacity allocation is developed in Section 3. The solution approach for the multi-objective model is covered in Section 4. Section 5 describes the optimization findings. Lastly, Sections 6 and 7 offer a discussion section and conclusion.

2. Architecture and Mathematical Modeling of Multi-Energy Microgrid Systems

Figure 1 depicts the connected cold–heat–electricity–hydrogen multi-energy microgrid system built in this work. The system structure description and mathematical model construction are shown below.

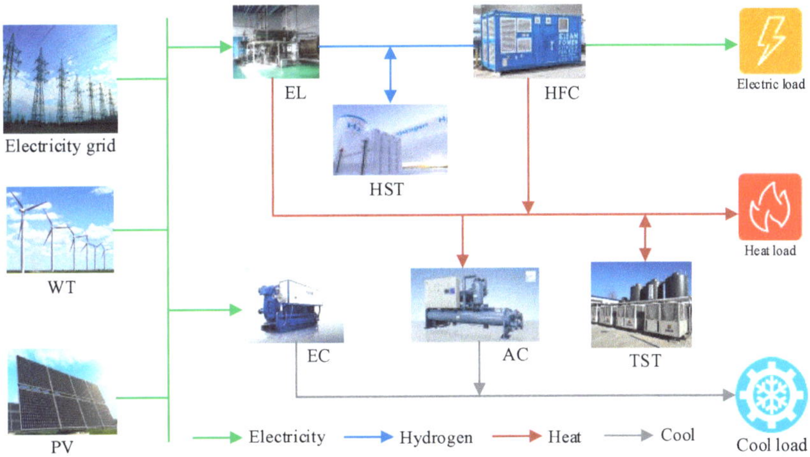

Figure 1. Multi-energy microgrid system architecture.

2.1. Multi-Energy Microgrid System Architecture

The photovoltaic (PV), wind turbine (WT), and upper grid connections provide the multi-energy microgrid system's electricity source. The system's primary energy conversion and storage components are the electric refrigerator (AR), thermal storage tank (TST), absorption chiller (AC), and a hydrogen energy storage system. Thermal, electrical, and cooling loads make up the system's loads.

Hydrogen storage systems include the EL, hydrogen storage tank (HST), and HFC. During the processes of electrolytic hydrogen production and HFC power generation, a significant amount of energy is dissipated as heat, resulting in low energy conversion efficiency for both EL and HFC. Therefore, utilizing the waste heat produced by HFC and EL is a crucial strategy for improving the overall energy conversion efficiency of the hydrogen storage system. The multi-energy coupling properties of EL and HFC may be fully realized by recovering the waste heat of EL and HFC through the waste heat recovery device. This also increases the system's efficiency in using energy and provides low-carbon clean electricity and heat energy [24].

2.2. Modeling of Equipment in Multi-Energy Microgrid Systems

The following displays the mathematical model for every device in the system.

2.2.1. Photovoltaic

The PV's output power can be obtained by measuring the present ambient temperature and the amount of solar energy that the panel has absorbed. The following formula expresses this relationship [25].

$$P_{PV}(t) = P_{sta} f_{PV} \frac{G(t)}{G_{sta}} \{1 + \alpha_T [T_a(t) - T_{a,sta}]\} \quad (1)$$

where $P_{PV}(t)$ is the output power of the PV panel; P_{sta} is the rated power of the PV under standard test conditions; f_{PV} is the attenuation coefficient of the power; $G(t)$ is the intensity of solar irradiation; G_{sta} is the intensity of solar irradiation under standard test conditions; α_T is the temperature coefficient of the PV; $T_a(t)$ is the temperature of the PV; and $T_{a,sta}$ is the ambient temperature under standard test conditions.

The temperature of the photovoltaic panel can be computed using the formula that follows.

$$T_a(t) = T(t) + \lambda G(t) \quad (2)$$

where $T(t)$ is the ambient temperature, and λ is the radiation temperature coefficient, which is 0.0256.

2.2.2. Wind Turbine

Wind energy can be converted by WT into electrical energy, and the wind speed at the unit's hub is what mostly influences its output power. For this reason, the formula for determining WT's output power is provided below [26].

$$P_{WT}(t) = \begin{cases} 0 & 0 < v < v_{in},\ v_r < v \\ \frac{P_r(v - v_{in})}{v_{out} - v_{in}} & v_{in} < v < v_{out} \\ P_r & v_{out} \leq v < v_r \end{cases} \quad (3)$$

where $P_{WT}(t)$ is the power generated by WT; P_r is the rated generating power of the WT; v_{in} is the cut-in wind speed of the WT; v_r is the rated wind speed of the WT; and v_{out} is the cut-out wind speed of the WT.

2.2.3. Electrolyzer

Water is converted to hydrogen and oxygen using EL, a type of hydrogen production apparatus. It is assumed in this study that the effectiveness of the EL is constant. Considering the thermoelectric coupling characteristics of the EL, the thermoelectric power relation equation for the EL is presented as follows [27].

$$\begin{cases} M_{el}(t) = \rho \eta_{el} P_{el,\,e}(t) \\ P_{el,\,h}(t) = (1 - \eta_{el}) P_{el,\,e}(t) \end{cases} \quad (4)$$

where $M_{el}(t)$ is the quantity of hydrogen that the EL produces; ρ is the quantity of hydrogen generated for 1 kW·h that the EL uses to generate energy; η_{el} is the EL's efficiency; the EL's input power is denoted by $P_{el,\,e}(t)$; and $P_{el,\,h}$ is the heat production power of the EL.

2.2.4. Hydrogen Fuel Cell

HFC uses hydrogen as a raw material and converts it into electricity through a redox reaction. Since HFC produces a lot of heat energy when producing electricity, waste heat recovery technology can be used to realize HFC cogeneration, which can increase HFC's energy consumption efficiency. Considering the thermoelectric effect of HFC, it is modeled using the following equation [27].

$$\begin{cases} P_{\text{hfc, e}}(t) = \frac{M_{\text{hfc}}(t)}{\rho}\eta_{\text{hfc}}, \\ P_{\text{hfc, h}}(t) = (1-\eta_{\text{hfc}})P_{\text{hfc, e}}(t) \end{cases} \quad (5)$$

where $P_{\text{hfc, e}}(t)$ is the power generated by HFC; $M_{\text{hfc}}(t)$ denotes the hydrogen consumption of the HFC; η_{hfc} is the HFC's efficiency; and $P_{\text{hfc, h}}(t)$ is the heat generation power of the HFC.

2.2.5. Hydrogen Storage Tank

The stable operation of hydrogen energy equipment in the microgrid system is dependent on the HST, which stores the hydrogen produced by the EL and supplies it to the HFC when required. The HST is modeled using the following equation [27].

$$M_{\text{hst}}(t) = M_{\text{hst}}(t-1) + \eta_{\text{hst}}M_{\text{el}}(t) - \frac{M_{\text{fc}}(t)}{\eta_{\text{hst}}} \quad (6)$$

where $M_{\text{ht}}(t)$ represents the amount of hydrogen stored in the tank at time t, and η_{hst} denotes the efficiency of HST.

2.2.6. Thermal Storage Tank

TST may efficiently increase the utilization rate of leftover heat from EL and HFC and prevent energy waste by storing heat when there is an excess of waste heat and releasing heat when there is a lack of waste heat. The modeling of TST follows a similar approach to that of HST, and its model is shown by the following formula [28].

$$H_{\text{tst}}(t) = (1-\eta_{\text{loss}})H_{\text{tst}}(t-1) + \eta^{+}H_{\text{cha}}(t) - \frac{H_{\text{dis}}(t)}{\eta^{-}} \quad (7)$$

where $H_{\text{tst}}(t)$ represents the TST's overall heat storage capacity; η_{loss} is the TST's dissipation coefficient; η^{+} is the heat charging and dissipation coefficient of the TST; $H_{\text{cha}}(t)$ A is the TST's heat charging capacity; $H_{\text{dis}}(t)$ is the TST's exothermic capacity; and η^{-} is the exothermic heat dissipation coefficient of the TST.

2.2.7. Electric Coolers

EC is an electric-cooling coupling device that converts electrical energy into cooling energy according to the system's requirements for cooling load. The model is shown below [29].

$$P_{\text{ec, c}}(t) = COP_{\text{ec}}P_{\text{ec, e}}(t) \quad (8)$$

where $P_{\text{ec, c}}(t)$ is the EC's cooling power; COP_{ec} is the rated coefficient of performance of the EC; and $P_{\text{ec, e}}(t)$ is the electrical energy required by the EC for cooling.

2.2.8. Absorption Chillers

AC can utilize the waste heat from EL and HFC as the driving energy source. It employs water, ammonia, or Freon as the refrigerant and uses a condenser and evaporator to enable these substances to absorb heat energy for refrigeration purposes. The model for AC is calculated by the subsequent formula [29].

$$P_{\text{ac, c}}(t) = COP_{\text{ac}}P_{\text{ac, h}}(t) \quad (9)$$

where $P_{\text{ac, c}}(t)$ is the cooling power of AC; COP_{ac} is the rated coefficient of performance of the AC; and $P_{\text{ac, h}}(t)$ is the thermal energy required for AC to achieve cooling.

3. Multi-Objective Capacity Optimization Allocation Model for Multi-Energy Microgrid Systems

The multi-objective capacity optimization allocation model of the system contains an economic objective function and an exergy objective function, as shown below.

3.1. Economic Objective Function

The primary components of multi-energy microgrids' economic costs are those associated with initial construction, operation and maintenance, and power purchasing. Therefore, the economic cost objective function can be expressed as follows [30,31].

$$f_1 = C_{\text{in}} \frac{r(1+r)^l}{(1+r)^l - 1} + C_{\text{om}} + C_{\text{gc}} \tag{10}$$

where C_{in} is the initial construction cost of multi-energy microgrid system; r is the inflation rate; l represents the system's design service life, set at 20 years in this study; C_{om} is the system's operating and maintenance costs; and C_{gc} is the cost of power purchased from the grid.

The following formula can be used to determine the system's initial construction cost.

$$C_{\text{in}} = k_{\text{pv}} N_{\text{pv}} + k_{\text{wt},} N_{\text{wt}} + k_{\text{hfc}} N_{\text{hfc}} + k_{\text{el}} N_{\text{el}} + k_{\text{ht}} N_{\text{ht}} + k_{\text{tst}} N_{\text{tst}} + k_{\text{ac}} N_{\text{e, ac}} + k_{\text{ec}} N_{\text{ec}} \tag{11}$$

where k_x is the power cost factor for each piece of equipment and N_x is the configured capacity of each piece of equipment.

The following formula can be used to determine the system's operating and maintenance cost.

$$\text{Com} = \sum_{t=1}^{T} (k_{\text{pv, om}} P_{\text{pv}}(t) + k_{\text{wt, om}} P_{\text{wt}}(t) + k_{\text{hfc, om}} P_{\text{hfc}}(t) + k_{\text{el, om}} P_{\text{el}}(t) + k_{\text{ac, om}} P_{\text{ac}}(t) + k_{\text{ec, om}} P_{\text{ec}}(t) + k_{\text{hst, om}} M_{\text{hst}}(t) + k_{\text{tst, om}} H_{\text{tst}}(t)) \tag{12}$$

where $k_{x,\text{ om}}$ is the cost of maintenance and operation for each equipment per unit of power and P_x represents the device's real operational power.

The following is the formula that the system uses to determine how much it will cost to acquire power from the grid.

$$C_{\text{gc}} = \sum_{t=1}^{T} \alpha_{\text{e}} P_{\text{e,buy}}(t) \tag{13}$$

where α_{e} is the power tariff that was purchased and $P_{\text{e,buy}}(t)$ is the system's purchased electricity.

3.2. Exergy Objective Function

Multi-energy microgrid systems must maximize energy utilization while also considering economic costs when making capacity allocations. Considering that a coupled cold–heat–electricity–hydrogen multi-energy microgrid system encompasses multiple types of energy and that the quality of energy can vary among these different types, it is important to recognize that energy quality is not necessarily uniform across all energy streams. As a result, evaluating the quality of any kind of energy source is essential. Exergy can evaluate the energy-saving indices of multi-energy microgrid systems in terms of both "quality" and "quantity". This approach is particularly applicable to multi-energy flow systems with multi-energy coupling, providing a reliable basis for the economic configuration and energy-saving operation of such systems.

The exergy of solar energy is considered to be the thermal radiation at the temperature of the Sun and is typically calculated using the Petela formula [32]. The photovoltaic panel's area, solar radiation intensity, ambient temperature, and solar temperature are all related to the exergy of solar energy, among other things. The expression for this exergy is shown in Equation (14).

$$E_{\mathrm{pv}}^{\mathrm{in}}(t) = I(t)S_{\mathrm{pv}}\gamma(t)\Delta t/1000 \tag{14}$$

$$\gamma(t) = 1 + \frac{1}{3}\left(\frac{T_{\mathrm{a}}(t)}{T_{\mathrm{sun}}}\right)^{4} - \frac{4}{3}\frac{T_{\mathrm{a}}(t)}{T_{\mathrm{sun}}} \tag{15}$$

where $I(t)$ is the solar irradiation intensity; S_{PV} is the area of the PV module; $T_{\mathrm{a}}(t)$ is the ambient temperature; and T_{sun} is the solar temperature, taken as 5777 K in this paper.

The physical nature of wind energy is kinetic energy; therefore, its exergy is related to factors such as wind speed and the area swept by the blades. The calculation equation is shown below [33].

$$E_{\mathrm{wt}}^{\mathrm{in}}(t) = \frac{8}{27}\rho A_{\mathrm{wt}} v_{\mathrm{wt}}^{3}(t) \tag{16}$$

where $v_{\mathrm{WT}}(t)$ is the wind speed; A_{WT} is the area swept by the wind through the blades of the fan, taken as 38.47 m^2; and ρ is the air density.

The power purchase exergy of the system is shown in Equation (17).

$$E_{\mathrm{e}}^{\mathrm{in}}(t) = P_{\mathrm{e,buy}}(t)\Delta t \tag{17}$$

where $P_{\mathrm{grid}}^{\mathrm{e}}(t)$ is the purchased power of the system and Δt is the operating period.

The exergy of the electrical load in the system is given by the following equation.

$$E_{\mathrm{e}}^{\mathrm{out}}(t) = P_{\mathrm{load,e}}(t)\Delta t \tag{18}$$

where $P_{\mathrm{load,e}}(t)$ represents the electrical load demand.

The following equation displays the exergy for the heat and cold loads.

$$\begin{cases} E_{\mathrm{h}}^{\mathrm{out}}(t) = (1 - \frac{T_{\mathrm{a}}}{T_{\mathrm{h}}})P_{\mathrm{load,\,h}}(t)\Delta t \\ E_{\mathrm{c}}^{\mathrm{out}}(t) = (\frac{T_{\mathrm{a}}}{T_{\mathrm{c}}} - 1)P_{\mathrm{load,\,c}}(t)\Delta t \end{cases} \tag{19}$$

where T_{a} is the ambient temperature, T_{h} is the heat source temperature, $P_{\mathrm{load,\,h}}(t)$ is the heat load demand, T_{c} is the cooling source temperature, and $P_{\mathrm{load,\,c}}(t)$ is the cooling load demand.

Exergy can effectively and accurately assess the overall energy efficiency of a multi-energy microgrid system. The exergy $E_{\mathrm{x}}^{\mathrm{in}}(t)$ input from the energy supply side of the microgrid system is introduced into the microgrid as payment exergy. It then undergoes a series of transmissions, conversions, and storage within the system, ultimately resulting in exergy $E_{\mathrm{x}}^{\mathrm{out}}(t)$ as output exergy. Payment exergy consists of $E_{\mathrm{pv}}^{\mathrm{in}}(t)$, $E_{\mathrm{wt}}^{\mathrm{in}}(t)$, and $E_{\mathrm{e}}^{\mathrm{in}}(t)$, and earnings exergy consists of $E_{\mathrm{e}}^{\mathrm{out}}(t)$, $E_{\mathrm{h}}^{\mathrm{out}}(t)$, and $E_{\mathrm{c}}^{\mathrm{out}}(t)$.

Exergy reflects the degree of matching between the system's energy supply and energy use at the energy level. The exergy of a multi-energy microgrid system can be calculated using the ratio of the system's input exergy to the output exergy. Thus, the system's exergy expression is as follows.

$$\varphi_e = \frac{E_{\mathrm{e}}^{\mathrm{out}}(t) + E_{\mathrm{h}}^{\mathrm{out}}(t)E_{\mathrm{v}}^{\mathrm{out}}(t)}{E_{\mathrm{e}}^{\mathrm{in}}(t) + E_{\mathrm{pv}}^{\mathrm{in}}(t) + E_{\mathrm{wt}}^{\mathrm{in}}(t)} \tag{20}$$

To facilitate a solution for the optimization model, the inverse of exergy is used as the system exergy objective function, as shown in the following equation.

$$f_2 = \frac{1}{\varphi_e} \tag{21}$$

3.3. Constraint

The power-balancing constraints, equipment operation constraints, and energy storage equipment operation limitations are the three primary categories of system constraints.

3.3.1. Power Balance Constraints

The power balance constraints contain electrical power balance constraints, thermal power balance constraints, and cold power balance constraints, as shown in Equations (22)–(24).

$$P_{pv} + P_{wt} + P_{e,buy} + P_{hfc,\,e} - P_{el,\,e} - P_{ec,\,e} = P_{load,e} \tag{22}$$

$$P_{hfc,\,h} + P_{el,\,h} - P_{ac,\,h} + H_{dis} - H_{cha} = P_{load,\,h} \tag{23}$$

$$P_{ac,\,c} + P_{ec,\,c} = P_{load,\,c} \tag{24}$$

3.3.2. Equipment Operating Constraints

The operating constraints of the WT, PV, and other equipment are displayed in Equation (25).

$$\begin{cases} 0 \le P_{pv}(t) \le N_{pv} \\ 0 \le P_{wt}(t) \le N_{wt} \\ 0 \le P_{el}(t) \le N_{el} \\ 0 \le P_{hfc}(t) \le N_{hfc} \\ 0 \le P_{ac}(t) \le N_{ac} \\ 0 \le P_{ec}(t) \le N_{ec} \end{cases} \tag{25}$$

Equation (26) illustrates the operating constraints of the two energy storage devices in the system, HST and TST.

$$\begin{cases} \alpha_{hst} N_{hst} \le M_{hst}(t) \le N_{hst} \\ M_{hst,\,start} = M_{hst,\,end} \\ \alpha_{tst} N_{tst} \le H_{tst}(t) \le N_{tst} \\ H_{tst,\,start} = H_{tst,\,end} \end{cases} \tag{26}$$

where α_{hst} and α_{tst} represent the minimum energy storage ratio coefficients for the hydrogen storage tank and thermal storage tank, respectively, with a value of 0.05; $M_{hst,\,start}$ and $M_{hst,\,end}$ denote the amount of hydrogen in the hydrogen storage tanks at the beginning and end of each day, respectively; N_{hst} and N_{tst} indicate the configuration capacities of the hydrogen storage tanks and thermal storage tanks, respectively; $H_{tst,\,start}$ and $H_{tst,\,end}$ represent the amount of heat stored in the thermal storage tanks at the beginning and end of each day, respectively.

4. Solution Strategies and Evaluation Methods for Multi-Objective Optimization Models

In this section, the solution scheme of the system capacity optimization allocation model developed in the previous section is investigated. The Pareto frontier of the model is solved using the ε-constraint method, and TOPSIS is used for decision-making. The multi-objective solution process is illustrated in Figure 2.

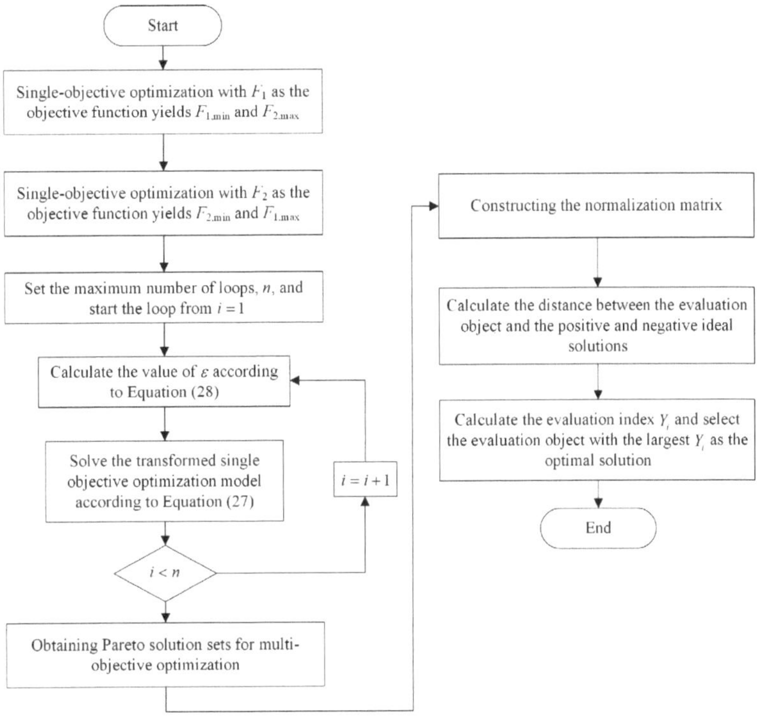

Figure 2. Multi-objective solution process.

4.1. Multi-Objective Solution Strategy

Traditional multi-objective optimization problems typically begin by transforming the multi-objective optimization model into a singular-objective one. This approach allows the multi-objective problem to be addressed by solving the resulting single-objective model. At the moment, the weighting approach and the ε-constraint method are the primary techniques for resolving multi-objective optimization issues [34]. The ε-constraint method is employed in this study to solve multi-objective optimization models due to its advantage of not requiring subjective weights and its ability to provide a complete set of solutions for the Pareto front.

The goal of the ε-constraint method is to convert an objective function in a multi-objective optimization model into a constraint. Then, by gradually altering the constraint's range of values, a multi-objective optimization problem can be converted into a sequence of single-objective optimization problems. This process is illustrated by the following equation [28,34].

$$\begin{cases} \min f_2 \\ \text{s.t.} \begin{cases} f_1 \leqslant \varepsilon \\ \text{eq.}(22) - \text{eq.}(26) \end{cases} \end{cases} \tag{27}$$

In Equation (27), ε represents the information gap parameter, and its value strategy is provided by Equation (28).

$$\varepsilon = f_{1,\max} - \frac{(f_{1,\max} - f_{1,\min})(i-1)}{i_{\max} - 1}, \; i = 1, 2, \cdots, n \tag{28}$$

where $f_{1,\max}$ denotes the economic cost of the system when optimized with exergy as the single objective; $f_{1,\min}$ represents the value of the objective function when the system is optimized with economic cost as the single objective; i indicates the number of cycles; and n is the maximum number of cycles.

The interval of ε ranges from the maximum to the minimum value of f_1. By varying ε, different objective function values can be obtained, collectively forming the Pareto optimal solution set for the multi-objective problem.

4.2. Evaluation Methodology

Selecting the best possible solutions for the multi-objective optimization issue is necessary because, with the Pareto solution set acquired by the ε-constraint technique, each and every point in the solution set represents the best possible solution. In this paper, the TOPSIS is used to make decisions regarding the points in the Pareto optimal solution set. TOPSIS is a technique for comprehensive evaluation that compares the distances between sample values and ideal solutions. First, the greatest and worst spots in the space are established as reference points. Next, a distance calculation is made between each evaluation object and these reference points. If an object is located closer to the best point and farther from the worst point, it is said to have superior comprehensive qualities [35].

The specific steps required to use TOPSIS for decision-making are listed in the following list.

1. Construct the normalized decision matrix.

$$f'_{ij} = \frac{f_{ij}}{\sqrt{\sum\limits_{i=1}^{n} f_{ij}^2}} \tag{29}$$

where f'_{ij} is the normalized objective value, and f_{ij} is the actual value of the i-th solution for the j-th objective.

2. Determine the positive ideal solution f^+ and the negative ideal solution f^-.

$$\begin{cases} f^+ = \{\max f_{ij} | j = 1, 2, \cdots, m\} \\ f^- = \{\min f_{ij} | j = 1, 2, \cdots, m\} \end{cases} \tag{30}$$

3. Calculate the Euclidean distance between each evaluation object and both the positive and negative ideal solutions.

$$\begin{cases} D_i^+ = \sqrt{\sum\limits_{j=1}^{m} (f'_{ij} - f_j'^+)^2} \\ D_i^- = \sqrt{\sum\limits_{j=1}^{m} (f'_{ij} - f_j'^-)^2} \end{cases} \tag{31}$$

4. Calculation of evaluation indicators.

$$Y_i = \frac{D_i^-}{D_i^- + D_i^+} \tag{32}$$

The evaluation index Y_i is introduced, with a larger value of Y_i indicating a better program.

5. Calculus Analysis

CPLEX can swiftly determine the optimization model's best solution and solve linear programming problems with good accuracy and resilience. Consequently, the multi-energy microgrid system's optimal configuration model is solved in this study by using the Yalmip toolbox (version 20220204) in MATLAB 2022R with the CPLEX 12.10; the instance settings and scenario analysis are displayed below.

5.1. Instance Setting

To validate the model proposed in this paper, meteorological and load data from a specific location are used as inputs to the system, as shown in Figures 3 and 4. The tariff for purchasing electricity from the grid for the multi-energy microgrid is presented in Table 1. The investment costs and parameters of the equipment in the system are detailed in Table 2. The key parameters of the equipment within the multi-energy microgrid system are presented in Table 3.

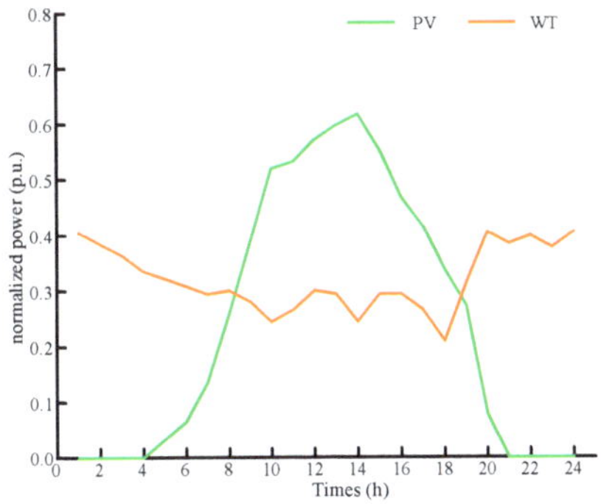

Figure 3. Wind and solar output factor.

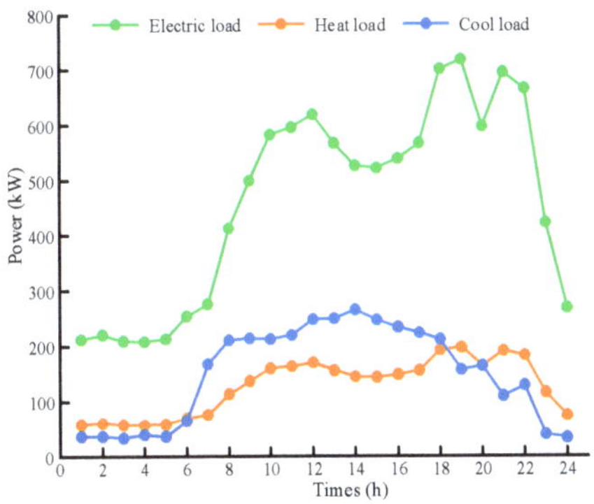

Figure 4. Heat and cool electrical loads in microgrid system.

This paper establishes four distinct scenarios for comparative analysis to examine the impact of the hydrogen energy storage system, considering the utilization of EL and HFC waste heat, on the capacity allocation outcomes of the multi-energy microgrid system, as well as on the multi-objective optimization of the system's performance.

Scenario 1: The capacity allocation is optimized with the best system economic cost using the microgrid model that is presented in this study in Section 2.1.

Scenario 2: In the microgrid model suggested in this study, the hydrogen storage system is replaced with battery electric storage (BES) and electric boiler (EB), and the capacity allocation is optimized with the best system economic cost.

Scenario 3: Capacity optimization with the objective of maximizing system exergy, using the microgrid model outlined in Section 2.1 of this paper.

Scenario 4: The microgrid model proposed in Section 2.1 is subjected to multi-objective optimization, considering both the economic cost and exergy of the system.

Table 1. Time-sharing tariff [36].

Times (h)	Price of Electricity (CNY/kW·h)
01:00–07:00 and 23:00–24:00	0.38
08:00–11:00 and 15:00–18:00	0.68
12:00–14:00 and 19:00–22:00	1.20

Table 2. Time-sharing tariff [29,30,36].

Equipment	Construction Cost (CNY)	Operation and Maintenance Cost (CNY)
PV	17,600	0.020
WT	12,000	0.070
EL	15,840	0.023
HFC	5000	0.060
HST	4928	0.014
TST	1760	0.141
EC	2200	0.160
AC	2502	0.152

Table 3. Key parameters of equipment within a multi-energy microgrid system [27,29,30,36,37].

Parament	Value	Parament	Value
f_{PV}	0.9	η_{el}	0.6
G_{sta}	1000 W/m2	η_{hfc}	0.6
α_T	$-3.7 \times 10-3$ °C	η_{hst}	0.9
$T_{a,sta}$	25 °C	η_{loss}	0.01
v_{in}	2.5 m/s	η^+, η^-	0.88
v_r	9.5 m/s	COP_{ec}	4.5
v_{out}	40 m/s	COP_{ac}	1.41

5.2. Analysis of Optimization Results for Scenario 1 and Scenario 2

The capacity optimization for Scenario 1 and Scenario 2 is conducted with the system economic cost as the objective. The results for the multi-energy microgrid system's cost and exergy are presented in Table 4, while the microgrid capacity configuration results are shown in Table 5.

As shown in Table 4, EL and HFC are used as thermoelectric coupling equipment in Scenario 1. The construction cost of Scenario 1 is CNY 1,485,800 higher than that of Scenario 2. Compared to Scenario 2, Scenario 1 has reduced operation and maintenance cost and power purchase costs by CNY 2,578,100 and CNY 532,800, respectively. The total cost of Scenario 1 is reduced by CNY 1,249,200 compared to Scenario 2, and exergy is improved by 0.22%.

Table 4. Optimization results for Scenario 1 and Scenario 2.

Item	Scenario 1	Scenario 2
Initial construction cost	6,244,200 CNY	4,758,400 CNY
Operation and maintenance cost	875,200 CNY	3,453,300 CNY
Power purchase cost	1,251,100 CNY	1,408,000 CNY
Total cost	8,370,500 CNY	9,619,700 CNY
Exergy	62.72%	61.91%

Table 5. Equipment capacity configuration results for Scenarios 1 and Scenario 2.

Equipment	Equipment Configuration Capacity	
	Scenario 1	Scenario 2
PV	1083.411 kW	1256.20 kW
WT	1376.79 kW	1191.40 kW
EL	808.15 kW	-
HFC	397.36 kW	-
HST	184.85 kW·h	-
TST	802.44 kW·h	84.87 kW
EC	300 kW	300 kW
AC	94.28 kW	94.28 kW
ESE	-	350.00 kW·h
EB	-	299.01 kW

As shown in Table 5, due to the discrepancy between the peak heat production hours of EL and HFC and the peak heat demand hours, excess heat produced by EL and HFC is stored in TST. This stored heat is then used during peak heat demand periods after the initial heat load demand has been met. Therefore, to ensure the reliability of the heat energy supply in the microgrid system, Scenario 1 requires larger TST configuration. In Scenario 2, the heat load demand is met by the electric boiler, which responds more quickly to heat load variations, thus necessitating a smaller configuration of TST.

5.3. *Analysis of Multi-Objective Optimization Result Analysis*

The results of the economic cost configuration and device capacities obtained from the optimization of scenarios 3 and 4 are presented in Tables 6 and 7.

Table 6. Optimization results for Scenario 3 and Scenario 4.

Item	Scenario 1	Scenario 2
Initial construction cost	6,244,200 CNY	4,758,400 CNY
Operation and maintenance cost	875,200 CNY	3,453,300 CNY
Power purchase cost	1,251,100 CNY	1,408,000 CNY
Total cost	8,370,500 CNY	9,619,700 CNY
Exergy	62.72%	61.91%

Table 7. Equipment capacity configuration results for Scenario 3 and Scenario 4.

Equipment	Equipment Configuration Capacity	
	Scenario 1	Scenario 2
PV	1083.411 kW	1256.20 kW
WT	1376.79 kW	1191.40 kW
EL	808.15 kW	-
HFC	397.36 kW	-
HST	184.85 kW·h	-
TST	802.44 kW·h	84.87 kW
EC	300 kW	300 kW
AC	94.28 kW	94.28 kW
ESE	-	350.00 kW·h
EB	-	299.01 kW

The total system cost for Scenario 4 increased by CNY 1,262,800 compared to Scenario 1, and its exergy improved by 3.81% relative to Scenario 1. Conversely, Scenario 4's total system cost is CNY 3,236,600 lower than that of Scenario 3, with a system exergy increase of 1.82% compared to Scenario 3.

As shown in Table 7, Scenario 3 focuses solely on optimizing system exergy. Consequently, the capacity configurations for each device obtained from Scenario 3 are higher compared to those from Scenario 1 and Scenario 4 optimizations. Compared to Scenario 1, Scenario 4 has a higher configured capacity for EL and HFC. This suggests that Scenario 4 effectively leverages the thermoelectric coupling characteristics of EL and HFC. The increased capacity of these components enhances the system's ability to utilize wind energy and reduces the cost of purchasing power for the microgrid.

By dividing a day into 24 one-hour time intervals, the electrical, thermal, and cooling energy balance diagrams for Scenario 4 are presented in Figures 5–7. Power values greater than 0 in the graph represent power generation or heat production, while power values less than 0 indicate power consumption or heat absorption.

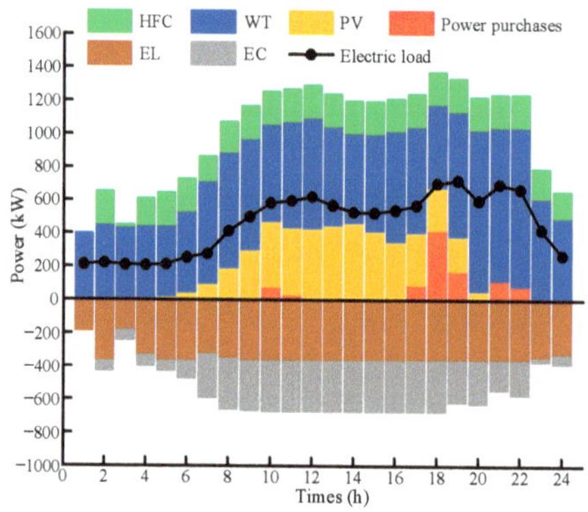

Figure 5. Power balancing in multi-energy microgrid system.

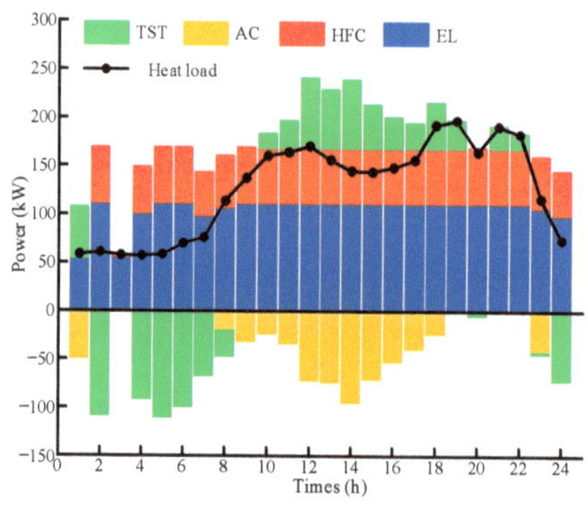

Figure 6. Thermal energy balance in multi-energy microgrid system.

As shown in Figure 5, the primary energy source for the microgrid system is the power generated by wind turbines and PV panels. The load demand peaks during the 10–11, 17–19, and 21–22 time periods. To maintain the stability of the system's energy supply, the system purchases power from the higher-level grid during these periods.

As shown in Figure 6, the heat load demand is low at moments 2 and 4–7. During these times, the excess heat energy, after meeting the heat load demand through EL and HFC, is stored by the TST. The heat load demand is high during the 18–19 and 21–22 time periods, when the TST releases thermal energy to compensate for the heat load deficit. As illustrated in Figure 7, the heat load demand is low, but the cooling load demand is high during the 13–15 time period. Consequently, during this period, the TST, along with EL and HFC, provides thermal energy to meet both the heat load and the cooling requirements of AC.

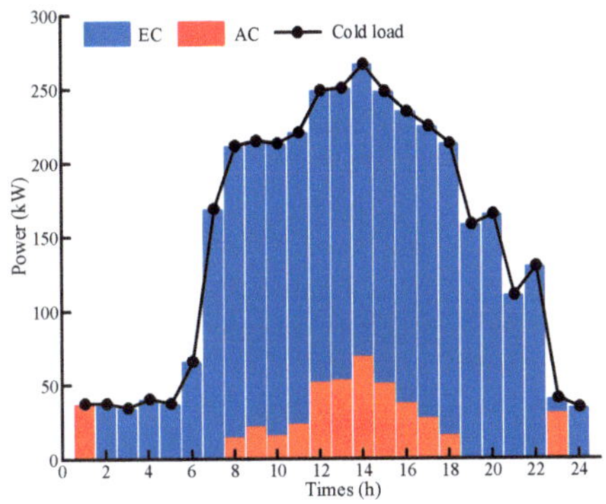

Figure 7. Cool energy balance in multi-energy microgrid system.

As shown in Figure 7, the cooling load demand in the system is met by both the EC and AC. The cooling load is higher during the 8–18 time period, during which the absorption chiller's power is utilized. The absorption chiller leverages low-grade thermal energy for cooling, in contrast to the EC, which uses high-quality electric energy. During these periods, the use of AC enhances the economic efficiency of energy supply and improves the system's exergy.

6. Discussion

In this research, a multi-energy microgrid system with combined cooling, heating, and power supply is realized by combining the operating waste heat of EL and HFC with EC and AC. Multi-objective capacity optimization is used to balance the microgrid system's energy efficiency and economic cost.

The multi-energy microgrid system built in Section 2.1 of this study has significantly increased economics and exergy efficiency when compared to the conventional microgrid system using BES and EG. The hydrogen storage system enhances the system's use of renewable energy sources and energy independence by converting excess electricity from wind power output into hydrogen energy that is stored in HST. This effectively controls the fluctuation of wind power generation. Thus, in comparison to the conventional microgrid system, the system built in this study has a CNY 532,800 lower power purchase cost, and a CNY 1,249,200 lower total economic cost. The exergy efficiency of the system built in this paper is improved by 0.22% when compared to the traditional microgrid system because the coupling of electricity–hydrogen–heat of hydrogen energy equipment is realized through the waste heat utilization of HFC and EL. This effectively improves the flexibility and reliability of energy supply on the load side.

Multi-objective optimization facilitates trade-offs and optimization among multiple objectives. A comparison of the results from multi-objective optimization with those from single-objective optimization reveals that single-objective optimization focused on economic cost can reduce the system's economic cost by CNY 1,622,800, albeit with a decrease in exergy efficiency of 3.81%. Conversely, optimizing solely for exergy efficiency results in an increase in economic cost by CNY 3,236,600 while improving exergy efficiency by 1.82%. Thus, economic cost and exergy efficiency represent conflicting optimization objectives. The optimal solution that accounts for both the economic cost and exergy efficiency of the microgrid system can be achieved through multi-objective optimization.

7. Conclusions

This paper presents the construction of a multi-energy microgrid system based on hydrogen energy with the goal of improving the system's energy supply reliability and resolving the conflict between the microgrid's economic cost and energy efficiency. With the goals of maximizing the system's energy efficiency and decreasing its economic cost, a multi-objective optimal allocation strategy for the multi-energy microgrid is presented. Ultimately, the solution is solved using TOPSIS and the ε-constraint approach. An arithmetic simulation was used to confirm the efficacy of the suggested approach, and the following findings were made:

1. The operating waste heat of EL and HFC is efficiently utilized by the multi-energy microgrid system built in this paper. The economic cost of the system suggested in this study is CNY 1,249,200 less than the conventional multi-energy microgrid system, and its exergy efficiency is increased by 0.22%.
2. In this paper, we employ multi-objective optimization to determine the optimal configuration of the microgrid system, considering both economic cost and exergy efficiency. The results show that multi-objective optimization enhances the exergy efficiency of the multi-energy microgrid by 3.81% compared to the exergy efficiency achieved through single-objective optimization focused solely on economic cost. Additionally, the economic cost from the multi-objective optimization is reduced by CNY 3,236,600 compared to that derived from single-objective optimization aimed only at maximizing exergy efficiency.

Although this research offers sufficient references for the optimization analysis of multi-energy microgrid systems, several elements still require additional research and development. The impact of unknown variables like wind power and load demand on the system configuration outcomes will be the first thing we look further into in future studies. Second, by carefully modeling the energy conversion components of the system and accounting for both the system's energy loss and the dynamic properties of the components, we will create a thorough energy management plan for the multi-energy microgrid. To further improve the multi-energy microgrid system described in this paper's practical practicality and application, we will also carry out a thorough analysis of the multi-energy microgrid's off-grid state configuration and operation techniques.

Author Contributions: Methodology, L.X.; formal analysis, Y.K.; investigation, W.L.; data curation, J.L.; writing—original draft, J.L.; writing—review and editing, Z.Y. and L.X.; supervision, Z.Y. and Y.W.; project administration, Y.W. All authors have read and agreed to the published version of the manuscript.

Funding: This research was funded by science and technology program of State Grid Xinjiang Power Co., Ltd. (B330DK240008).

Data Availability Statement: The original contributions presented in the study are included in the article, further inquiries can be directed to the corresponding authors.

Conflicts of Interest: Authors Ji Li, Lei Xu, Yang Kou, Weile Liang, and Yunshan Wang were employed by the Electric Power Research Institute of State Grid Xinjiang Electric Power Co., Ltd. The remaining author declares that the research was conducted in the absence of any commercial or financial relationships that could be construed as potential conflicts of interest. The Electric Power Research Institute of State Grid Xinjiang Electric Power Co., Ltd., had no role in the design of the study; in the collection, analyses, or interpretation of data; in the writing of the manuscript; or in the decision to publish the results.

Nomenclature

Abbreviation	Full Name
HFC	hydrogen fuel cell
EL	electrolyzer
PV	photovoltaic
WT	wind turbine
TST	thermal storage tank
EC	electric chiller
AC	absorption chiller
HST	hydrogen storage tank
BES	battery electric storage
EB	electric boiler

References

1. George, A.R.; Al-Hinai, A.; Margaret, V. A comprehensive review on energy management strategy of microgrids. *Energy Rep.* **2023**, *9*, 5565–5591.
2. Verma, N.; Kumar, N.; Gupta, S.; Malik, H.; Márquez, F.P.G. Review of sub-synchronous interaction in wind integrated power systems: Classification, challenges, and mitigation techniques. *Prot. Control. Mod. Power Syst.* **2023**, *8*, 1–26. [CrossRef]
3. Liu, L.; Yao, X.; Qi, X.; Han, Y. Low-carbon economy configuration strategy of electro-thermal hybrid shared energy storage in multiple multi-energy microgrids considering power to gas and carbon capture system. *J. Clean. Prod.* **2023**, *428*, 139366. [CrossRef]
4. Shen, Y.; Hu, W.; Liu, M.; Yang, F.; Kong, X. Energy storage optimization method for microgrid considering multi-energy coupling demand response. *J. Energy Storage* **2022**, *45*, 103521. [CrossRef]
5. Lu, Z.; Gao, Y.; Xu, C.; Li, Y. Configuration optimization of an off-grid multi-energy microgrid based on modified NSGA-II and order relation-TODIM considering uncertainties of renewable energy and load. *J. Clean. Prod.* **2023**, *383*, 135312. [CrossRef]
6. Chen, J.J.; Qi, B.X.; Rong, Z.K.; Peng, K.; Zhao, Y.L.; Zhang, X.H. Multi-energy coordinated microgrid scheduling with integrated demand response for flexibility improvement. *Energy* **2021**, *217*, 119387. [CrossRef]
7. Chen, H.; Yang, S.; Chen, J.; Wang, X.; Li, Y.; Shui, S.; Yu, H. Low-carbon environment-friendly economic optimal scheduling of multi-energy microgrid with integrated demand response considering waste heat utilization. *J. Clean. Prod.* **2024**, *450*, 141415. [CrossRef]
8. Bhandari, R.; Adhikari, N. A comprehensive review on the role of hydrogen in renewable energy systems. *Int. J. Hydrogen Energy* **2024**, *82*, 923–951. [CrossRef]
9. Qiu, Y.; Li, Q.; Ai, Y.; Chen, W.; Benbouzid, M.; Liu, S.; Gao, F. Two-stage distributionally robust optimization-based coordinated scheduling of integrated energy system with electricity-hydrogen hybrid energy storage. *Prot. Control. Mod. Power Syst.* **2023**, *8*, 1–14. [CrossRef]
10. Salama, H.S.; Magdy, G.; Bakeer, A.; Vokony, I. Adaptive coordination control strategy of renewable energy sources, hydrogen production unit, and fuel cell for frequency regulation of a hybrid distributed power system. *Prot. Control. Mod. Power Syst.* **2022**, *7*, 1–18. [CrossRef]
11. Li, J.; Xiao, Y.; Lu, S. Optimal configuration of multi microgrid electric hydrogen hybrid energy storage capacity based on distributed robustness. *J. Energy Storage* **2024**, *76*, 109762. [CrossRef]
12. Nguyen, Q.M.; Nguyen, D.L.; Nguyen, Q.A.; Pham, T.N.; Phan, Q.T.; Tran, M.H. A Bi-level optimization for the planning of microgrid with the integration of hydrogen energy storage. *Int. J. Hydrogen Energy* **2024**, *63*, 967–974. [CrossRef]
13. Nojavan, S.; Akbari-Dibavar, A.; Farahmand-Zahed, A.; Zare, K. Risk-constrained scheduling of a CHP-based microgrid including hydrogen energy storage using robust optimization approach. *Int. J. Hydrogen Energy* **2020**, *45*, 32269–32284. [CrossRef]
14. Fang, X.; Dong, W.; Wang, Y.; Yang, Q. Multiple time-scale energy management strategy for a hydrogen-based multi-energy microgrid. *Appl. Energy* **2022**, *328*, 120195. [CrossRef]
15. Shi, M.; Wang, H.; Lyu, C.; Xie, P.; Xu, Z.; Jia, Y. A hybrid model of energy scheduling for integrated multi-energy microgrid with hydrogen and heat storage system. *Energy Rep.* **2021**, *7*, 357–368. [CrossRef]
16. Li, H.; Liang, F.; Guo, P.; He, C.; Li, S.; Zhou, S.; Deng, L.; Bai, C.; Zhang, X.; Zhang, G. Study on the biomass-based SOFC and ground source heat pump coupling cogeneration system. *Appl. Therm. Eng.* **2024**, *165*, 114527. [CrossRef]
17. Zhang, Y.; Zhu, N.; Zhao, X.; Luo, Z.; Hu, P.; Lei, F. Energy performance and enviroeconomic analysis of a novel PV-MCHP-TEG system. *Energy* **2023**, *274*, 127342. [CrossRef]
18. Kim, J.; Han, C.; Lee, S.; Kim, Y. Performance analysis of PEMFC-assisted renewable energy source heat pumps as a novel active system for plus energy buildings. *Energy Build.* **2024**, *319*, 114477. [CrossRef]
19. Yuan, J.; Xiao, F.; Gang, W.; Zhang, Y.; Shi, J.; Zhang, Z.; Hao, X. Load allocation methods for the thermal and electrical chillers in distributed energy systems for system efficiency improvement. *Energy Convers. Manag.* **2023**, *292*, 117334. [CrossRef]
20. Tong, C.; Zhang, H.; Liu, H.; Tian, W.; Wang, C. Study on energy efficiency improvement strategies of photovoltaic-hybrid energy storage DC microgrids under the concept of green energy conservation. *Appl. Math. Nonlinear Sci.* **2024**, *9*, 1–20. [CrossRef]

21. Meng, Z.; Wang, K.; Di, J.; Lang, Z.; He, Q. Energy analysis and exergy analysis study of a novel high-efficiency wind-hydrogen storage and power generation polygeneration system. *Int. J. Hydrogen Energy* **2024**, *57*, 338–355. [CrossRef]
22. Yang, S.; Wang, G.; Liu, Z.; Deng, C.; Xie, N. Energy, exergy and exergo-economic analysis of a novel SOFC based CHP system integrated with organic Rankine cycle and biomass co-gasification. *Int. J. Hydrogen Energy* **2024**, *53*, 1155–1169. [CrossRef]
23. Xu, A.; Yang, L.; Huang, W.; Zhang, Y.; Long, H.; Liu, Z.; Deng, C.; Yang, S. Exergy, economic, exergoeconomic and environmental (4E) analyses and multi-objective optimization of a PEMFC system for coalbed methane recovery. *Energy Convers. Manag.* **2023**, *297*, 117734. [CrossRef]
24. Yang, C.; Dong, X.; Wang, G.; Lv, D.; Gu, R.; Lei, Y. Low-carbon economic dispatch of integrated energy system with CCS-P2G-CHP. *Energy Rep.* **2024**, *12*, 42–51. [CrossRef]
25. Koholé, Y.W.; Ngouleu, C.A.W.; Fohagui, F.C.V.; Tchuen, G. Optimization of an off-grid hybrid photovoltaic/wind/diesel/fuel cell system for residential applications power generation employing evolutionary algorithms. *Renew. Energy* **2024**, *224*, 120131. [CrossRef]
26. Das, B.K.; Tushar, M.S.H.K.; Hassan, R. Techno-economic optimisation of stand-alone hybrid renewable energy systems for concurrently meeting electric and heating demand. *Sustain. Cities Soc.* **2021**, *68*, 102763. [CrossRef]
27. Sun, L.; Chen, L.; Xiong, Y.; Li, X.; Mei, W. Capacity optimization configuration of hydrogen energy storage cogeneration integrated energy system considering photothermal collector module. *Electr. Power Autom. Equip.* **2023**, *43*, 70–76.
28. Cao, Y.; Wang, Q.; Du, J.; Nojavan, S.; Jermsittiparsert, K.; Ghadimi, N. Optimal operation of CCHP and renewable generation-based energy hub considering environmental perspective: An epsilon constraint and fuzzy methods. *Sustain. Energy Grids Netw.* **2019**, *20*, 100274. [CrossRef]
29. Ma, X.; Liu, S.; Zhao, S.; Zong, Q.; Liu, H. The optimal configuration of distributed generators for CCHP micro-grid based on double-layer operation strategy and improved NSGA-III algorithm. *Energy Build.* **2023**, *293*, 113182. [CrossRef]
30. Han, Y.; Yu, S.C.; Li, L.Y.; Hou, Y.; Li, Q.; Chen, W. Low-carbon and economic configuration method for solar hydrogen storage microgrid including stepped carbon trading. *High Volt. Eng.* **2022**, *48*, 2523–2533.
31. Zhang, J.; Kan, S.; Hua, Y.; Zhang, X.; You, G.; Zhang, Z. Capacity Optimization of CHP Microgrid Based on Hydrogen Energy Storage. *Acta Energ Sol Sin.* **2022**, *43*, 428–434.
32. Chen, Y.; Guo, M.; Liu, Y.; Wang, D.; Zhuang, Z.; Quan, M. Energy, exergy, and economic analysis of a centralized solar and biogas hybrid heating system for rural areas. *Energy Convers. Manag.* **2023**, *276*, 116591. [CrossRef]
33. Ehyaei, M.A.; Ahmadi, A.; Rosen, M.A. Energy, exergy, economic and advanced and extended exergy analyses of a wind turbine. *Energy Convers. Manag.* **2019**, *183*, 369–381. [CrossRef]
34. Mohebi, P.; Roshandel, R. Optimal design and operation of solar energy system with heat storage for agricultural greenhouse heating. *Energy Convers. Manag. X* **2023**, *18*, 100353. [CrossRef]
35. Wang, X.; Henshaw, P.; Ting, D.S.K. Three-objective Optimization of a Concentrated Photovoltaic Thermoelectric System via Student Psychology-based Optimization Algorithm and an External Archive Strategy. *Energy* **2024**, *304*, 131915. [CrossRef]
36. Wu, Q.; Li, C. Modeling and operation optimization of hydrogen-based integrated energy system with refined power-to-gas and carbon-capture-storage technologies under carbon trading. *Energy* **2023**, *270*, 126832. [CrossRef]
37. Chen, Y.; Chen, J.; Zhang, W.; Ni, C.; Zhao, B. Optimal scheduling strategy of distributed electric-thermo-hydrogen system considering lifetime decay characteristics of electrolytic cell and battery. *Electr. Power Autom. Equip./Dianli Zidonghua Shebei* **2023**, *43*.

electronics

MDPI

Review

Hydrogen Energy in Electrical Power Systems: A Review and Future Outlook

Siting Dai [1], Pin Shen [1], Wenyang Deng [2] and Qing Yu [3,*]

[1] Taipa Campus, City University of Macau, Macao SAR 999078, China; f21092100239@cityu.edu.mo (S.D.); f23092100403@cityu.edu.mo (P.S.)
[2] School of Electric Power, South China University of Technology, Guangzhou 510640, China; dwyang@scut.edu.cn
[3] Management College, Guangzhou City University of Technology, Guangzhou 510800, China
* Correspondence: yuqing@gcu.edu.cn

Abstract: Hydrogen energy, as a zero-carbon emission type of energy, is playing a significant role in the development of future electricity power systems. Coordinated operation of hydrogen and electricity will change the direction and shape of energy utilization in the power grid. To address the evolving power system and promote sustainable hydrogen energy development, this paper initially examines hydrogen preparation and storage techniques, summarizes current research and development challenges, and introduces several key technologies for hydrogen energy application in power systems. These include hydrogen electrification technology, hydrogen-based medium- and long-term energy storage, and hydrogen auxiliary services. This paper also analyzes several typical modes of hydrogen–electricity coupling. Finally, the future development direction of hydrogen energy in power systems is discussed, focusing on key issues such as cost, storage, and optimization.

Keywords: hydrogen energy; electric–hydrogen coupling; electrolytic hydrogen production; fuel cells; future power systems

check for updates

Citation: Dai, S.; Deng, W.; Shen, P.; Yu, Q. Hydrogen Energy in Electrical Power Systems: A Review and Future Outlook. *Electronics* **2024**, *13*, 3370. https://doi.org/10.3390/electronics13173370

Academic Editor: Carlos Andrés García-Vázquez

Received: 26 July 2024
Revised: 14 August 2024
Accepted: 15 August 2024
Published: 25 August 2024

1. Introduction

The continued growth of the world's population and urbanization rate has significantly increased energy demand and carbon emissions [1]. To reduce carbon emissions and preserve energy supply, countries worldwide are actively investigating ways to decrease fossil energy consumption and increase the use of renewable energy sources in the energy system [2]. In China, carbon emissions from the power sector account for nearly 40% of the total carbon dioxide emissions of the whole of society [3]. Integrating renewable energy sources is critical to achieving low-carbon operation of the power system and mitigating man-made climate change.

Natural factors can lead to intermittency and uncertainty in renewable energy generation. Some large megacities, such as Guangzhou, Shanghai, Tokyo, and New York, face challenges in managing a diverse range of new energy sources and addressing uneven load distribution. During low-demand periods when many renewable energy sources are connected to the grid, there is a problem of excess power, which can severely affect the safe and reliable operation of the power system [4]. The use of storage batteries to store excess renewable energies will effectively avoid the high penetration rate of renewable energy access to the grid caused by the grid pressure [5]. However, traditional energy storage, mainly lithium compounds and lead-acid batteries, generally have limited by cycle life, unsatisfactory charging and discharging efficiency, and self-discharge problems [6].

Through continuous exploration by researchers, hydrogen has emerged as an energy carrier with great potential. Hydrogen-powered electricity generation has zero carbon emissions and only water as a byproduct, thus not contributing to air pollution [7]. One of the key benefits of hydrogen is its high energy density. For the same mass, hydrogen can provide three times more energy than gasoline when burned [8], and it can be derived from a variety of sources, including water, oil, natural gas, biofuels, and even sewage sludge [9,10]. A detailed comparison table is included below as Table 1. The table provides a comprehensive overview of how hydrogen fares against traditional fuels, highlighting its potential as a viable alternative for a low-carbon future.

Table 1. Comparison of energy content and energy density of hydrogen with other fossil fuels [3–10].

Fuel	Ratio of Energy and Mass (MJ/kg)	
	Higher Heating Value	Lower Heating Value
Hydrogen	142	120
Methane	55.5	50
Liquefied Petroleum Gas	50.0	46.1
Gasoline	46.4	44
Natural Gas	55	50
Coal (anthracite)	24	35
Ethanol	29.7	26.8

By producing hydrogen through electrolysis for power storage, renewable energy sources can be used more flexibly and their negative impact on the power grid can be reduced. The introduction of electrolyzed hydrogen into fuel cells will enable the development of distributed power sources, stand-alone power generation, and co-generation facilities. In addition, hydrogen electrification based on fuel cells will allow electricity to be produced, when and where it is needed, greatly increasing the flexibility of power generation and reducing losses. Therefore, research on key technologies for hydrogen energy production and storage, electrification, and hydrogen–electricity coupling will be an important part of building a new low-carbon power system.

This paper initially examines the current state of research on hydrogen preparation and storage technologies. It subsequently categorizes the essential technologies for hydrogen electrification and outlines typical approaches to hydrogen–electricity integration. Finally, it explores future directions for applying hydrogen energy, offering insights for future research in electricity power systems.

2. Current Status of Hydrogen Production and Storage

This section will review hydrogen electrolytic preparation and storage technologies, including the principles of electrolytic hydrogen production, preparation methods, electrolyzer technology, and several major types of hydrogen storage.

2.1. Review of Production of Hydrogen from Electrolytic Water

Industrial hydrogen production requires stable gas output, cost control, and simple preparation technology. The main industrial hydrogen production methods include hydrogen from fossil fuels, hydrogen from coke oven gas, hydrogen from methanol, etc. [11,12].

The process of utilizing natural gas, carbon dioxide and methane cracking to produce hydrogen has a long history. Under high-temperature and pressured environments with catalysts, by reacting with water vapour, these gases produce a variety of gases including hydrogen; coke oven gas is a by-product of the coke-making process, mainly including hydrogen and methane [13]. Hydrogen purification involves separating various gases from coke oven gas using solid adsorbents. Methanol reacts with water vapour at specific temperatures and pressures to yield hydrogen and carbon dioxide, which are subsequently separated via pressure swing adsorption to achieve increased hydrogen purity [14].

Although the above methods are relatively mature, they consume high energy in the preparation process and rely on fossil fuels, which will produce carbon-containing waste gas, and this is not conducive to reducing carbon emissions.

By running a direct current through a tank containing electrolytes, water can be broken down to produce oxygen and hydrogen. This process is known in the industry as electrolytic hydrogen production. The significant advantages of this method are that the raw materials are easy to obtain, the process is relatively simple, and no carbon emissions are generated during the preparation process, which reflects the environmental friendliness. However, it is worth noting that the electrolytic water process requires a large amount of electrical energy consumption, which is a major challenge of the method [15].

Alternative methods for hydrogen production, such as ethanol and saccharide reforming, bio-photolysis of water, photochemical water splitting, and high-temperature water splitting, are currently in the developmental phase and exhibit limited technological maturity [16]. Nowadays, as new energy technologies are increasingly deployed on a broad scale, the cost of electricity from photovoltaics, wind turbines, and similar sources continues to decline. This trend has sparked growing interest in the electrolytic production of hydrogen from water. The following section will introduce the principle of electrolysis of water and the development of electrolysis technology in detail.

(a) Technology for electrolyzing water

In an electrolytic water system, two electrodes are immersed in an electrolyte solution and linked to a power supply to enable current flow [17]. When a sufficient DC voltage is applied across the electrodes, water molecules undergo decomposition, yielding hydrogen at the cathode and oxygen at the anode. The electrolyte, introduced into the solution, enhances water conductivity and supports uninterrupted electron movement. Typically, the electrolyte is selected from acidic or alkaline substances and utilizes various ions as charge carriers, such as H^+, OH^-, O^{2-}, etc. [18].

To keep the electrolysis reaction controllable, most commercial electrolyzers use current control; the hydrogen production rate in this case can be fixed at a set current value [19].

The efficiency of a hydrogen electrolysis system relates to the voltage efficiency of the cell bank of the electrolysis unit and the efficiency of the associated auxiliary equipment when it is in operation [20]. The efficiency of an electrolytic water system can be quantified by the ratio of the fuel's high heating value (HHV) to the electrical energy input, expressed as Equation (1) [21]. In addition to the commonly used high calorific value efficiency, voltage efficiency is also one of the important indicators to evaluate the performance of water electrolysis systems [22]. The calculation formula of voltage efficiency is shown in Equation (2), and its level directly affects the energy consumption and cost of the electrolytic process. In addition, there are Faraday efficiency, thermal efficiency, and overall system efficiency, which provide an important perspective for a comprehensive understanding of the efficiency of hydrogen electrolysis systems [23–25]. The calculation of the Faraday efficiency is based on the comparison of the actual amount of hydrogen generated with the theoretical amount of hydrogen generated and is crucial for evaluating the chemical selectivity of the electrolysis process [24]. Thermal efficiency is concerned with the thermodynamic efficiency of energy conversion during electrolysis [25]. The calculation of thermal efficiency usually involves a comparison between the input electrical energy and all the thermal energy losses involved in the electrolysis process (such as electrolyte preheating, electrolytic cell heat dissipation, etc.), which is of great significance for optimizing the energy balance of the system [26]. Overall system efficiency is a comprehensive evaluation index that takes into account the efficiency of the electrolysis unit and its auxiliary equipment. It not only includes the voltage efficiency and Faraday efficiency of the electrolyzer itself, but also involves the efficiency of the power supply system (such as photovoltaic panels, wind turbines, and other renewable energy equipment), losses in the transmission and distribution of electrical energy, and the storage and transportation efficiency after electrolytic water hydrogen production. The calculation of the overall energy efficiency of the system is complex and requires consideration of multiple variables, but it provides a comprehensive perspective for assessing the economic and environmental sustainability of the entire hydrogen electrolysis system.

$$\eta_{EL} = \frac{\text{HHV}\left(\frac{\text{kWh}}{\text{kg}}\right) \times \text{Hydrogen production (kg)}}{\left[\frac{\text{Battery power input (kWh)}}{\text{Energy supply efficiency}}\right] + \text{Energy ancillary loss (kWh)}} \quad (1)$$

$$Voltage\ Efficiency = \frac{Theoretical\ resolution\ voltage}{Cell\ Voltage} \cdot 100\% \quad (2)$$

(b) Electrolyzer technology

Currently, common types of water electrolysis plants used for hydrogen production include alkaline water electrolyzers (AWEs), polymer electrolyte membrane electrolyzers (PEMEs), and solid oxide electrolyzers (SOEs) [26]. Among them, AWEs have a longer history of development and mature technology and dominate the electrolyzer market, while SOEs have a higher operating temperature, lower power consumption, and higher efficiency during electrolysis [27,28]. PEMEs have been developing rapidly in recent years and use proton exchange membranes instead of diaphragms and electrolytes in traditional electrolyzers. The three main types of electrolyzers and their characteristics are summarized in Table 2.

Table 2. Three main parameters of electrolytic cell technology [26–31].

Parameters	AWEs	PEMEs	SOEs
Technology maturity	Widespread commercialization	commercialization	R&D phase
Temperature/°C	60–80	50–80	600–1000
Pressure/bar	10–30	20–50	1–15
Current density/A·cm^{-2}	<0.45	1.0–2.0	0.3–1.0
Individual electric power/MW	6	2	0.15
Electrical efficiency/%	62–82	67–82	81–86
System energy consumption/kWh·m^{-1}	4.2–4.8	4.4–5.0	2.5–3.5
Area of electric stack/m^2	3–3.6	<0.13	<0.06
Hydrogen production rate/m^3·h^{-1}	1400	400	<10
Reactor life/kh	55–120	60–100	8–20
System life/a	20–30	10–20	-
Hydrogen production purity %	>99.8	99.999	-
Cold start time/min	60–120	5–10	>60
Hot start time/s	60–300	<10	900
Investment cost/USD·kW^{-1}	800–1500	400–2100	>2000

Electrolyzers are usually composed of multiple cell banks connected in series. This configuration allows for the cumulative voltage to increase significantly, even though each cell operates at a relatively low voltage of approximately 2 volts. Moreover, at high current densities, electrolyzers with parallel-connected banks of cells can be scaled to achieve megawatt outputs at relatively low voltages (up to a few kVs) [28]. For smooth operation of the system, a power supply unit and other energy supply auxiliaries are also required, as shown in Figure 1: Water is pumped into the electrolytic cell and heated by a heat exchanger to reach the operating temperature [29]. The power supply unit incorporates a transformer and rectifier to supply direct power to the electrolyzer. Following electrolysis, gases generated from water enter a gas separator for initial separation from water, followed by purification and drying processes [30].

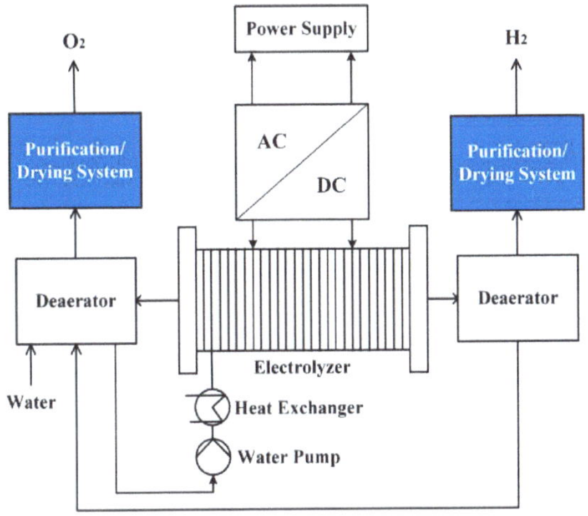

Figure 1. Electrolyzer structure.

2.2. Developments in Hydrogen Storage Technologies

For the past several years, there has been rapid growth in renewable energy generation. However, these energy sources are usually highly unstable and time- and seasonally dependent, severely affecting power system tides. To better consume renewable energy sources, reliable energy storage methods need to be developed to deal with the volatility and stochasticity of renewable energy sources. Common energy storage methods include pumped storage, compressed air, and chemical batteries [31–35]. Compared with them, based on the function that can smooth out the volatility and uncertainty of new energy sources, hydrogen has the advantages of a strong energy storage capacity, long storage time, and high flexibility, which can be used to realize the following functions [36,37].

Consumption time shift: When supply exceeds demand, hydrogen balances demand and supply by storing excess energy generated by renewable energy sources When demand increases, hydrogen can be used to generate electricity directly or inject electricity into the grid via fuel cells. Especially during times of reduced demand and lower electricity rates, energy can be stored directly in hydrogen to minimize energy expenses; during peak demand periods and when electricity prices are elevated, hydrogen is used to generate electricity to maximize returns. In addition, hydrogen has a much longer effective storage time of weeks or even months compared to batteries, which have an effective storage time of only a few hours or weeks [38].

Seasonal change response: Due to seasonal differences in new energy production, hydrogen can be used to transfer renewable energy sources across seasons. In addition, due to its high energy density, hydrogen can reach MWh or even TWh of storage capacity compared to the kWh to MWh storage capacity of chemical batteries [39,40].

Hydrogen storage and electrochemical energy storage, represented by lithium compound batteries, are two principal energy storage technologies. In terms of energy density, gaseous hydrogen has an energy density of approximately 33.6 kWh/kg, while compressed gaseous hydrogen at 700 bar has a volumetric energy density of about 1.7 kWh/L [41]. Lithium compound batteries have a maximum energy density of up to 0.3 kWh/kg, with a volumetric energy density of approximately 0.75 kWh/L [42]. Regarding cycle life, hydrogen fuel cells typically have a maximum cycle life of no more than 10,000 h, whereas lithium compound batteries generally offer around 2000 charge–discharge cycles, with each cycle experiencing a capacity loss of about 10–20%. In terms of cost, fuel cell power generation based on low-cost grey hydrogen is approximately USD 0.1–$0.2 per kWh, while the cost of green hydrogen produced from renewable energy ranges from USD 0.2 to USD 0.5 per kWh [43]. The generation cost for lithium compound batteries is around USD 0.1–USD

0.2 per kWh, with the potential to decrease to USD 0.05–USD 0.1 per kWh in large-scale grid storage projects due to economies of scale and technological advancements [44,45]. Additionally, hydrogen fuel cells generally have a start-up time on the order of minutes, contrasting with the second-scale response of lithium batteries. In terms of volume, lithium compound batteries are several cubic metres depending on requirements, while fuel cells typically have a volume of no less than several cubic metres [46,47].

From the above comparison, it is evident that hydrogen storage has significant advantages in energy density and service life, while lithium batteries excel in rapid start-up and device volume. Consequently, hydrogen storage is more suitable for seasonal and large-scale energy storage applications, such as addressing grid balancing and the intermittency of renewable energy sources, as well as applications requiring long endurance and high energy density, such as heavy-duty trucks, long-haul shipping, and aviation. On the other hand, electrochemical energy storage represented by lithium compound batteries is well suited for rapid-response applications, such as electric vehicles, grid frequency regulation, and home energy storage systems, as well as for power supply in portable devices such as mobile phones and laptops.

As mentioned above, the development of hydrogen storage technology is an essential prerequisite for the construction of hydrogen-containing power systems. Conventional solutions usually involve storing hydrogen in the form of compressed gases and cryogenic liquids; for extensive applications, underground hydrogen storage has been demonstrated to be an advantageous approach. In recent years, solid-state hydrogen storage has developed rapidly and is considered to be the safest mode of hydrogen storage [48].

To further understand the feasibility and practicality of various hydrogen storage and transportation methods, a comprehensive evaluation of their pros, cons, costs, and overall evaluations is essential. Table 3 summarizes the key aspects of several forms of hydrogen storage and transportation, based on the criteria mentioned.

Table 3. The advantages, disadvantages and cost evaluation of different hydrogen storage and transportation methods [41–48].

Category	Pros	Cons	Cost	Comment
Compressed Gaseous Hydrogen	Mature, suitable for short-distance transportation and distribution	The storage density is relatively low, requiring high-pressure containers, which limits the transportation volume	Short-distance transportation (within 200 km): USD 0.30 to USD 0.50 per kg of H_2. Long-distance transportation: USD 1.00 to USD 3.00 per kg of H_2	Suitable for localized or short-distance applications, but the costs for long-distance transportation are higher.
Liquid Hydrogen	High storage density, suitable for large-scale, long-distance transportation	The liquefaction process has high energy consumption, strict low-temperature storage with transportation requirements, and may have evaporation loss	Medium distances (around 500 km): USD 1.00–USD 2.00/kg H_2	Liquid hydrogen is suitable for long-distance transportation, but the high cost of the liquefaction process and cryogenic storage equipment has a significant impact on the overall cost. High storage density makes it suitable for long-distance transportation but requires considering the energy consumption during the liquefaction process.

Table 3. *Cont.*

Category	Pros	Cons	Cost	Comment
Chemical Hydrogen Carriers	Easy to store and transport at room temperature and pressure, with large transportation volume and high safety	Requires additional hydrogen extraction processes, affecting the overall efficiency	Depending on the chemical carrier used, the cost is between USD 1.00 and USD 5.00/kg H_2. e.g., the transport cost of using ammonia as a carrier is about USD 1.50–USD 3.00/kg H_2, and the additional cost of hydrogen extraction from the carrier should be considered.	Suitable for long-distance or cross-border transportation, safer at normal temperature and pressure, but the overall efficiency is low.
Metal Hydrides	High hydrogen storage density, stable storage, suitable for small-scale or portable applications	The hydrogen storage/release process is slow, and the cost of metal hydride material is high	USD 5.00–USD 10.00/kg H_2, mainly due to high material cost and low efficiency of hydrogen storage/release process.	High storage density, but the process of releasing hydrogen is complex and usually used for special applications rather than large-scale transportation.
Solid Hydrogen	Storage density is extremely high, theoretically	The preparation and processing techniques are complex with high cost, and mainly in the lab stage	At present, it is mainly in the research stage, and the economy has not been fully evaluated. If the practical application is considered, the cost may be extremely high, far more than other forms.	Suitable for experiments and specific applications; commercial and large-scale applications are not yet available.

(a) Compressed Gaseous Hydrogen

Compressed gaseous hydrogen is a common method for hydrogen storage and transportation. By compressing hydrogen gas to high pressures, its volume is significantly reduced, making storage and transportation more economical and efficient. Containers for storing compressed hydrogen are typically made from high-strength materials, such as carbon fibre composites, to withstand pressures exceeding 700 bar [49]. Compressed hydrogen can be transported via specially designed high-pressure hydrogen transport vehicles, pressurized pipeline systems, or dedicated trains. As a key form of hydrogen storage technology, compressed gaseous hydrogen benefits from a high gravimetric energy density (approximately 33.6 kWh/kg) and mature technology, making it widely applicable in fields such as transportation and energy storage [50].

However, there are several challenges associated with compressed gaseous hydrogen. Firstly, high-pressure storage requires robust containers, such as high-pressure cylinders, which significantly increase equipment and maintenance costs [51]. Secondly, the compression process is energy-intensive, potentially reducing overall energy efficiency. Additionally, while hydrogen has a high gravimetric energy density, its volumetric energy density is relatively low, which can be a limiting factor in applications requiring high volumetric energy density. The flammability and explosiveness of hydrogen also necessitate stringent safety measures, adding complexity to storage and transportation [52,53].

Future development directions for compressed gaseous hydrogen technology include enhancing compression efficiency and reducing storage costs. This involves developing new, efficient compression technologies and materials to lower the overall economic burden, as well as integrating with renewable energy sources (such as wind and solar power) to reduce carbon emissions [54].

(b) Liquid Hydrogen

Hydrogen can be liquefied by cooling it to extremely low temperatures, approximately −253 °C (20.28 K). Liquid hydrogen, stored in insulated tanks, can be transported overland [54]. As an efficient method for hydrogen storage and transportation, liquid hydrogen presents significant advantages in current technology and applications [55]. However, several challenges constrain its development, like energy efficiency (liquefying hydrogen requires approximately 30–40% of the hydrogen's energy to operate the refrigeration equipment, which severely impacts economic viability) [56]; boil-off losses (even with high-efficiency insulation materials, liquid hydrogen will evaporate at a rate of 0.1–0.3% per day during storage and transportation [57]. This gradual evaporation means that long-term storage can result in significant hydrogen losses, particularly during long-distance transport, which affects overall economics); cost problems (the costs of cryogenic storage tanks, liquefaction equipment, and transportation infrastructure are high [58]. These costs can substantially increase the unit cost of liquid hydrogen, especially in small-scale applications); and safety concerns (the extremely low temperature of liquid hydrogen poses risks of personal injury to operators, and the highly flammable nature of hydrogen increases the risk of fire or explosion. Furthermore, the ageing of insulation materials can exacerbate boil-off losses and safety risks) [59].

Future research on liquid hydrogen will focus on developing more efficient liquefaction technologies to reduce energy consumption [60]. For instance, employing advanced heat exchange technologies and more efficient compressors or utilizing renewable energy sources to drive the liquefaction process could significantly lower production costs. The development of new and high-efficiency insulation materials, such as those based on nanotechnology, could further reduce evaporative losses during storage. These materials would more effectively block heat transfer, enhancing the thermal efficiency of storage tanks [61]. Expanding hydrogen infrastructure to facilitate the large-scale application of liquid hydrogen will also contribute to reducing its cost.

(c) Chemical Hydrogen Carriers

Hydrogen can be stored and transported in the form of chemical compounds through chemical reactions. Compounds that can react with hydrogen include ammonia (NH_3), methanol (CH_3OH), and liquid organic hydrogen carriers (LOHCs) such as N-ethyl carbazole and dimethyl toluene [62]. Compared to compressed hydrogen gas, chemical hydrogen carriers can store hydrogen at a higher volumetric energy density under ambient temperature and pressure. This form of hydrogen has a higher energy density and lower storage pressure requirements, which helps to reduce storage and transportation costs. Many chemical hydrogen carriers have a long history of commercial use, such as ammonia in the fertilizer industry, indicating relatively mature technologies [63,64].

There are several challenges associated with chemical hydrogen carriers. Firstly, hydrogenation and dehydrogenation processes for chemical carriers like ammonia often involve complex and energy-intensive procedures, requiring sophisticated equipment [65]. The processing of chemical hydrogen carriers may produce by-products or waste, necessitating extra treatment steps and potentially increasing overall costs [66]. Handling ammonia and certain hydrogen compounds can also involve environmental and safety concerns, such as the toxicity of ammonia and the high-temperature stability issues of hydrogen compounds [67]. Moreover, the efficiency and economic viability of hydrogen recovery and reuse remain critical issues that need continuous optimization of related technologies and processes.

Research on chemical hydrogen carriers will focus on several key areas [68–70]: firstly, improving the efficiency of hydrogen dehydrogenation and hydrogenation processes and developing low-energy, cost-effective chemical hydrogen carrier technologies to reduce costs; secondly, enhancing the safety of chemical hydrogen carriers to minimize environmental and health impacts; and finally, investigating and developing new chemical carriers to improve their advantages in terms of energy density, storage conditions, and processing costs.

(d) Metal Hydrides

Under specific conditions, hydrogen can react with certain metals or metal alloys to form hydrides, which can release hydrogen gas under certain conditions. Magnesium hydride (MgH_2) and sodium borohydride ($NaBH_4$) are notable for their high hydrogen storage density [71]. Metal hydrides are significant in hydrogen storage and transportation due to their high hydrogen storage density and relatively low safety risks. Their high gravimetric energy density and lower storage pressure requirements enable hydrogen to be stored in solid form at ambient or moderate temperatures, thereby reducing the safety risks associated with high-pressure storage [72,73]. The reactions involved in storing and releasing hydrogen with metal hydrides are reversible, allowing for efficient recovery and reuse of hydrogen.

There are still some mountains researchers need to conquer associated with metal hydrides. Firstly, the hydrogen release process typically requires high temperatures, which can lead to higher energy consumption [74]. Secondly, the hydrogen storage capacity of certain metal hydrides may be limited by material stability and cycle life in practical applications, potentially causing performance degradation over time [75]. Furthermore, the production and handling costs of metal hydrides are relatively high, requiring precise control and advanced technology during preparation, which can increase the overall economic burden [76].

As a result, researchers are focusing on several key areas: optimizing the synthesis and processing of hydrides to improve hydrogen adsorption and release efficiency, reduce energy consumption, and lower production costs; developing new metal hydride materials to enhance hydrogen storage capacity and cycle stability; and investigating and optimizing the operational temperature ranges of hydrides to reduce energy consumption and improve the economic viability of practical applications.

(e) Solid Hydrides

Under extremely low temperatures and high pressures, hydrogen can crystallize to form solid hydrogen, which can be stored in specialized materials [77]. Solid hydrogen materials allow for storage at lower pressures, reducing the need for high-pressure tanks, and they typically offer higher hydrogen density, thereby enhancing the energy density of hydrogen storage [78]. Solid hydrogen storage systems generally exhibit good long-term stability and lower leakage risks, making them safer compared to liquid hydrogen and high-pressure gaseous hydrogen.

However, solid hydrogen also presents several challenges. The cost of storage materials is relatively high, particularly for rare metal hydrides [79]. Moreover, the processes of hydrogen release and adsorption often require higher temperatures and pressures, which can increase the system's energy consumption and complexity. The reaction rates and hydrogen release rates of solid hydrogen storage materials may also limit their efficiency and practicality in certain applications [80]. The material's cycle life and hydrogen recovery efficiency are important factors that need to be addressed.

Researchers are focused on developing new types of solid hydrogen materials, seeking more cost-effective solutions, and optimizing the processes of hydrogen adsorption and release [81].

2.3. *Application of Integration of Hydrogen Storage with Renewable Energy Sources*

The integration of hydrogen storage with renewable energy sources is a crucial pathway to achieving sustainable energy systems. Hydrogen storage technologies enable the conversion of intermittent renewable energy production, such as wind and solar power, into reliable and adjustable hydrogen energy storage. Hydrogen energy and renewable energy have been applied in various industrial scenarios [82–86]. The American company

SunHydrogen has developed a photocatalytic technology that utilizes semiconducting materials with high photocatalytic performance to enhance hydrogen production efficiency, aligning hydrogen production closely with solar energy generation to achieve a sustainable hydrogen supplement [82]. The Norwegian Hywind project employs floating wind turbines to enable hydrogen production and storage under variable wind conditions [83]. In Southern Australia, the Hydrogen Park project utilizes electricity generated from hydropower stations for hydrogen production and storage, demonstrating the practical integration of hydropower with hydrogen energy [84]. In addition, Europe has achieved the integration of hydrogen energy with biomass energy. These application cases illustrate various models of hydrogen storage technology within renewable energy systems, advancing technological development and providing valuable practical experience and demonstration effects for future energy systems [85,86].

3. Key Technologies for Hydrogen Electrification

Hydrogen energy is usually connected to the power system through an electrification process as an energy carrier; the electrification of hydrogen is usually realized in the form of gas-to-electricity conversion to release energy. In this section, hydrogen electrification technologies based on fuel cells will be introduced, and medium- and long-term storage methods and ancillary services based on hydrogen energy will also be summarized.

3.1. Hydrogen Electrification Technology

Fuel cells can maximize the energy contained in hydrogen and can convert the chemical energy of hydrogen directly into electricity with an efficiency of 60–80% and only water as a by-product of the reaction [87].

(a) Principle of operation

As shown in Figure 2, a hydrogen fuel cell delivers hydrogen to the anode, initiating ionization to release electrons and H^+ ions, while air is supplied to the cathode to generate negative oxygen ions [88]. Like an electrolyzer, various fuel cell types vary in their charge transfer direction and electrolyte charge carriers, potentially producing the water on either side. Typically employing platinum-coated carbon materials as catalysts, different fuel cell types interact with diverse electrolytes. Table 4 provides an overview of these electrolytes' characteristics. Later, an example of a polymer electrolyte membrane fuel cell (PEMFC) will be discussed.

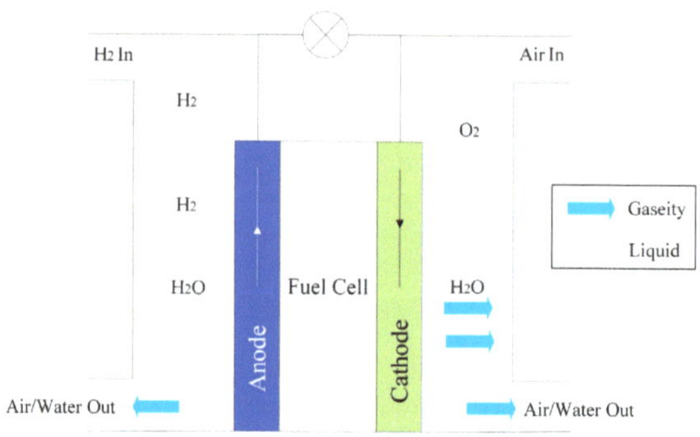

Figure 2. Fuel cell working principles.

Table 4. Operating temperature, cell voltage efficiency and advantages and disadvantages of different types of fuel cells [89–94].

Typology	Operating Temperature	Battery Pack Voltage Efficiency	Advantage	Drawbacks
Proton Exchange Membrane Fuel Cell (PEMFC)	80 °C–100 °C (low temperature) or 200 °C (high temperature)	50–60%	Fast start-up and versatility	Catalysts are expensive
Solid Oxide Fuel Cells (SOFCs)	800 °C–1000 °C	60–80%	With solid electrolytes, the reaction heat is reusable and less costly.	Presence of metal corrosion problems
Alkaline fuel cell (AFC)	Approx 70 °C	About 60%	Good current response	Limited application scenarios
Molten Carbonate Fuel Cell (MCFC)	Approx 650 °C	60–80%	Good conductivity and high current density	Slow start, only for large-scale use.
Phosphoric acid fuel cell (PAFC)	Approx 180 °C	More than 80%	High efficiency	Low current density and high catalyst cost

(b) Structure of Fuel Cell Systems

A typical fuel cell system consists of a battery pack and its auxiliary equipment, including hydrogen tanks, pumps, air compressors, power electronics, thermal management systems, etc., as shown in Figure 3. A single fuel cell can produce a rated voltage of 0.6 V to 0.8 V at rated loads [95], and the voltage of the battery pack can be boosted by increasing the number of cells. Similar to the electrolyzer, paralleling fuel cell packs can increase the current and thus the output capacity. In addition, the current can be increased by increasing the effective area of the cells. Typically, fuel cell systems are also equipped with a number of auxiliary devices, wherein a compressor is used to feed air into the reaction tank, and a hydrogen storage tank can output hydrogen to the reaction tank at a controlled flow rate and pressure. In addition, the fuel cell system typically includes a condenser to cool the incoming compressed air, alongside a humidification device that prevents the proton exchange membrane in the reactor tank from drying out [96]. Typically, an inverter is also installed to convert the generated DC power to AC power.

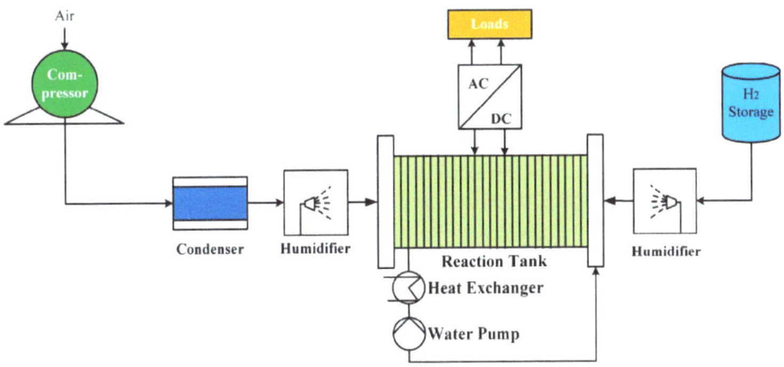

Figure 3. Fuel cell system structure.

Typically, the highest efficiency of a fuel cell is attained when supplying power to the load. Lowering the current density below its peak power density can mitigate cell voltage losses, thereby enhancing cell efficiency [97]. Systematically, fuel cells can be engineered to operate within the optimal efficiency range through strategic system control and design [98,99].

3.2. Hydrogen Ancillary Service Technologies

Hydrogen energy systems, comprising both fuel cells and electrolyzers, offer versatile ancillary services to the grid. These include peak load management, frequency regulation, mitigation of negative electricity pricing occurrences, voltage stabilization, and emergency power restoration efforts [100].

Similar to other forms of energy storage, hydrogen can initially address issues related to power supply constraints in transmission and distribution lines caused by inadequate capacity [101]. The cost of scheduling to address the "congestion" of power flow can be significantly higher. To solve the "congestion" problem of tidal currents, the dispatch costs increase considerably. Flexible application of the electrolysis–electrification process of hydrogen will allow efficient and low-cost regulation of grid currents.

Frequency regulation aims to uphold grid stability by closely aligning grid frequency and amplitude with their respective reference values. This is achieved through coordinated injection or absorption of power, ensuring equilibrium between electricity supply and demand. In current power systems, frequency regulation typically operates with two tiers. The initial tier involves primary frequency control, which ensures ongoing management during brief deviations from the nominal frequency [102]. For frequency deviations of more than 30 s, a second level, frequency restoration control, is required. This level of control has a greater capacity to provide frequency regulation over longer time scales. Fuel cells and electrolyzers can implement these two levels of frequency control, which can increase or decrease the reference value of power output based on a frequency signal [102]. During instances of frequency reduction, the fuel cell can ramp up power generation, while the electrolyzer can adjust electrolysis rates downward. This capability makes it a cost-effective device enabling efficient demand response.

Helping to reduce the negative prices that occur in the electricity market is a major advantage of electrified hydrogen equipment. Negative electricity prices occur primarily due to a lack of flexibility on the generation side. In electricity markets with high hydrogen penetration, a similar but more flexible regulation of the spot price of electricity can be achieved by increasing or decreasing the rate of hydrogen production in the electrolyzer, or by adjusting the output power of the fuel cell, similar to that of a generator.

Hydrogen devices offer an additional capability known as voltage support [103]. These devices, linked to the grid through power electronic converters, can regulate the power factor of both the fuel cell and electrolyzer. This regulation involves adjusting their power outputs to meet the voltage requirements at the grid connection point. In the event of a power outage, the fuel cell allows for a noiseless and fast black start compared to a conventional generator [104].

3.3. Hydrogen–Electric Power Systems

Like gasoline engines, certain internal combustion engines or turbines can operate directly using hydrogen, generating power as a result. However, owing to hydrogen's comparatively lower volumetric energy density, the thermodynamic efficiency of hydrogen internal combustion engines typically ranges from 20% to 25%, which is less than that of gasoline internal combustion engines [105]. In addition, although no carbon dioxide is released, the combustion of hydrogen produces nitrogen dioxide as an air pollutant [106]. The use of fuel cells as a power source in a hydrogen–electric power system will be effective in avoiding air pollution.

Vehicles equipped with fuel cells will differ from conventional new energy vehicles in that the driving distance is limited due to insufficient battery capacity; the former will have a longer driving distance. It is expected that 3% of cars sold globally in 2030 will be fueled by hydrogen, and by 2050, the percentage could reach 36% [107]. Numerous companies are actively advancing fuel cell powertrain technology, focusing on enhancing both reliability and safety. Toyota's Mirai fuel cell vehicle, for instance, utilizes a commercially available PEMFC with a volumetric power density of 3.1 km/L and a peak power output of 144 kW. Future advancements aim to enhance the safety and compactness of hydrogen storage in vehicles. Presently, the majority of commercial hydrogen fuel cell vehicles utilize high-pressure compressed hydrogen fuel tanks for storage.

In recent years, there has been notable progress in the development of fuel-cell ships alongside fuel-cell vehicles. Ship emissions during navigation contribute approximately 2.5% to global greenhouse gas emissions. Fuel cells offer sufficient power for ships travelling long distances and can fulfil the auxiliary energy requirements of large vessels, contrasting with battery-powered ships. This trend is also observed in fuel cell trains, with hydrogen-powered regional trains already operational in Europe and projected to capture over 30% of the market in the future [108,109].

4. Key Technologies for Hydrogen–Electric Coupling

Electric–hydrogen coupling systems have been studied by scholars, and it has been demonstrated that considering generation and transmission planning within the framework of electric–hydrogen integration will reduce the total system cost [110]. In general, electric–hydrogen coupling can be categorized according to the location of implementation, including load-side and power-side electric–hydrogen coupling.

4.1. Load-Side Electric–Hydrogen Coupling

The load-side electricity–hydrogen coupling model is shown in Figure 4 [111]. Taking China as the example, electricity from large-scale centralized renewable energy power plant stations in the northwestern part of China can be transported to the load-intensive central and eastern parts of China through extra-high-voltage transmission lines, and then hydrogen can be produced through electrolysis on the load side to achieve voltage support or capacity reserve for the power grid through hydrogen fuel cells or hydrogen gas turbines.

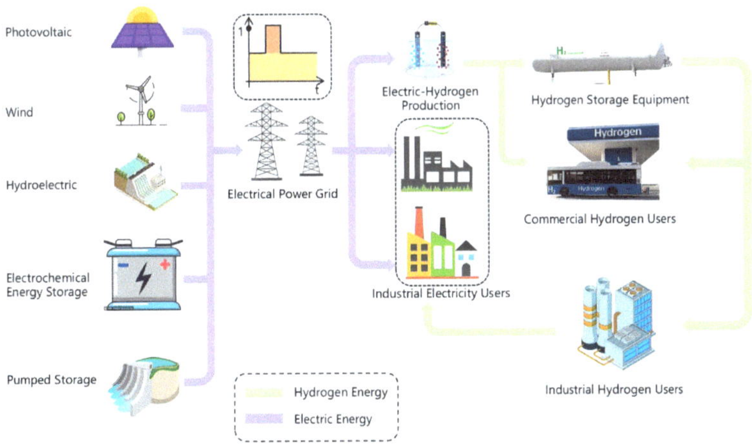

Figure 4. Schematic diagram of load-side electric–hydrogen coupling mode.

The load-side electricity–hydrogen coupling model has several advantages. First, hydrogen production plants can be flexibly located. Given that the electricity powering the electrolyzer originates from the grid, hydrogen plants can be strategically located near regions with significant hydrogen demand. This approach reduces expenses associated with hydrogen conversion, storage, and transporting over long distances, thereby mitigating potential safety hazards linked to extensive, long-range hydrogen logistics.

Secondly, UHV (ultra-high-voltage) transmission lines can be fully utilized for optimal allocation of large-scale renewable energy. As indicated by the research findings, the expense of hydrogen pipeline transport rises from USD 4.5/kg to USD 10/kg with a distance increase from 400 km to 1000 km [112,113]. Conversely, for the identical distance, the expense of electricity transmission via UHV lines escalates from USD 0.05/kWh to USD 0.09/kWh. In terms of energy equivalence, the unit cost for transmitting electricity is approximately one-fourth to one-fifth of that for hydrogen conveyance. Therefore, concerning energy distribution, UHV lines present a more favourable economic competitiveness compared to hydrogen pipelines.

Finally, there are large differences in peak and valley electricity prices near load centres. This can provide a degree of financial compensation for losses in the electricity–hydrogen–electricity energy conversion process.

Nevertheless, this approach carries certain drawbacks. Primarily, the initial lower capital outlay for UHV lines does not invariably translate into reduced overall energy transmission expenses. Given the intermittent nature of renewable energy sources, they must be complemented with coal, hydroelectricity, electrochemical storage, and pumped storage to achieve a more consistent power output for UHV lines. This integration necessitates additional investments in flexible power sources and storage systems, potentially driving up the overall system costs.

The operational flexibility of electrolyzers might not be fully utilized, as the variability of renewable energy has already been mitigated through adaptable power sources and storage batteries. The relative stability of power transmitted to the load side limits the performance of electrolyzers in providing grid balancing services and absorbing renewable energy fluctuations on the power side [114].

4.2. Power-Side Electric–Hydrogen Coupling

Electricity–hydrogen coupling modes on the power supply side include three categories: transmission mode, hydrogen transmission mode, and local balancing mode [115].

(a) Electricity-transmission method

Figure 5 illustrates this conceptual framework [116]. Within this setup, electrolyzers, hydrogen storage facilities, and hydrogen gas turbines are strategically positioned on the power side to mitigate losses associated with renewable energy sources. The electrolysis tanks exhibit robust responsiveness, effectively harnessing power derived from renewable sources; the resultant hydrogen serves dual purposes, catering to local consumption or facilitating re-electrification.

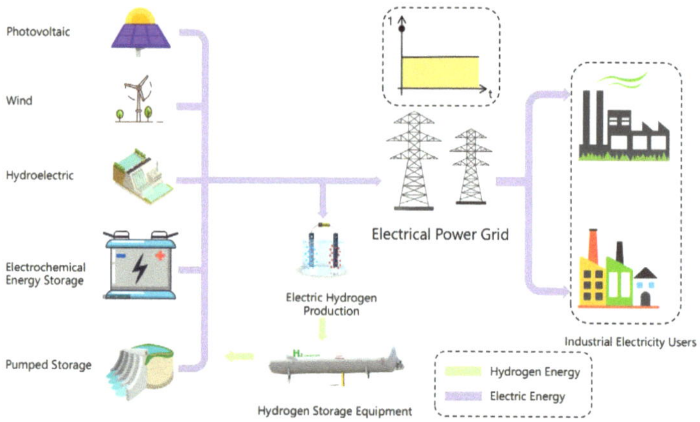

Figure 5. Schematic diagram of power-side electric–hydrogen coupling—transmission mode.

This model offers significant benefits by optimizing both renewable energy integration and transmission line efficiency. As the number of coal and natural gas power plants gradually decreases, electrolyzers, hydrogen storage systems, and hydrogen gas turbines play a crucial role in peak shaving and grid balancing services. During periods of high renewable energy output, electrolysis tanks operate at maximum capacity, storing hydrogen in dedicated tanks. Conversely, when renewable energy production dips, hydrogen gas turbines engage to stabilize transmission lines, maximizing renewable energy utilization while ensuring high operational efficiency.

Similar to the load-side electric–hydrogen coupling model, this model has an inefficient electric–hydrogen–electric conversion process with high energy losses.

(b) Hydrogen-delivery mode

Figure 6 illustrates the schematic representation of the model [117]. Renewable energy-derived hydrogen generated at the power generation site is transported via pipelines to the central load centre located at a considerable distance.

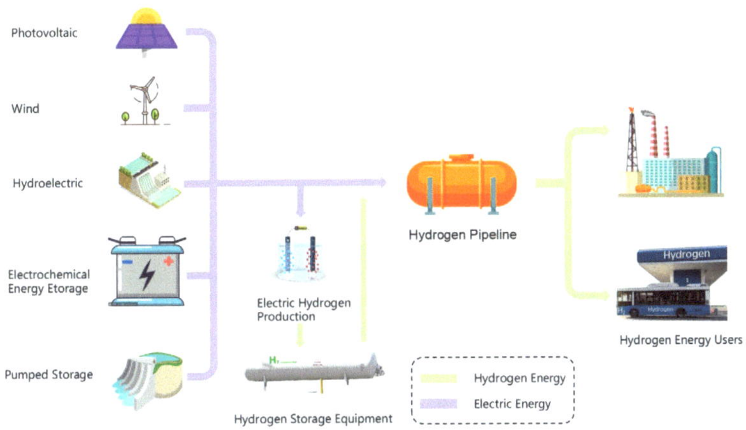

Figure 6. Schematic diagram of remote electric–hydrogen coordinative hydrogen delivery mode.

The advantage of this model is that the transmission of renewable energy over long distances through hydrogen reduces the pressure on the power grid and can effectively support the large-scale development and utilization of renewable energy. Unlike electricity, the supply and demand for hydrogen do not need to be balanced instantaneously and are therefore more tolerant of fluctuations. Also, pipelines can be used as a means of storage.

However, the construction of hydrogen pipelines remains expensive and is in its initial phases, posing constraints on hydrogen transmission. Conventional hydrogen trucking is viable only for limited distances and small-scale operations. Although blending hydrogen into natural gas pipelines can lower costs, the blend ratio should not surpass 10%, thus constraining large-scale hydrogen transport.

(c) Local equilibrium

This model utilizes locally produced hydrogen from renewable energy sources on the power side. Its primary strengths include cost-effectiveness and enhanced efficiency. By avoiding the need for hydrogen electrification, energy losses during conversion processes are minimized. However, the model's effectiveness heavily relies on local hydrogen demand. Insufficient demand necessitates transporting hydrogen generated from renewable sources to other regions. Consequently, industries reliant on hydrogen as raw material or fuel, such as chemical production, steel manufacturing, and fuel-cell vehicles, should strategically locate themselves in regions abundant in renewable energy resources [118].

5. Outlook for the Development of Hydrogen Energy Applications

Given the advancements in hydrogen production, transportation, storage, and integration with electricity, along with supportive policies promoting hydrogen as a pivotal energy source, its research and application are poised for substantial growth. This paper consolidates existing research findings on hydrogen energy, projecting future developments in the power sector to emphasize the following key areas:

5.1. Reducing Electrolysis Costs and Improving Preparation Yields

The primary expense in producing hydrogen via water electrolysis is the electricity cost, directly impacting the overall electrolysis cost. Consequently, optimizing energy consumption stands as the critical factor in reducing the production cost of hydrogen through water electrolysis. There are two main ways to reduce the cost: the first is to further develop the electrolysis membrane technology to reduce the energy consumption in the electrolysis process; the second is to use cheaper renewable electricity.

For electrolytic membrane technology, different technology routes appear to be advancing together; although the AWE is more mature, it also faces the problem of service life. Therefore, research is needed to increase the long-term stability of AWEs by improving the chemical, mechanical, and thermal stability of the membranes, as well as by using highly conductive polymer compositions to increase ionic conductivity.

PEM technology presents distinct advantages compared to alkaline electrolysis, including elevated operational current density, superior gas purity, increased outlet pressure, and reduced spatial requirements. Nonetheless, the primary obstacle facing this technology remains the cost associated with its components. Reducing membrane thickness by enhancing mechanical resistance can improve efficiency and durability, and reducing the loading of noble materials by adjusting the surface properties of the catalyst material (e.g., increasing the surface area) can improve the kinetic properties of the electrode material, which in turn reduces the power consumption of the PEM.

The SOE is an efficient technology that is rapidly evolving. However, its main challenge is durability. Enhancing electrolyte conductivity and refining the chemical and mechanical robustness of the electrolyzer is crucial for enhancing durability. Moreover, adjusting the electrochemical surface characteristics and ensuring compatibility with the electrode materials can extend the lifetime of SOEs.

5.2. Development of New Materials to Enhance Hydrogen Storage Capacity

Currently, high-pressure storage stands out as the most viable and extensively employed hydrogen storage technique in transportation applications. However, it comes with inherent drawbacks. Firstly, due to the high pressures involved, storage tanks necessitate costly and challenging-to-produce high-strength materials. Secondly, compressed hydrogen occupies a considerable volume, restricting storage capacity per unit space. Thirdly, elevated pressures heighten the risk of leaks or ruptures, posing safety concerns. Hence, innovative materials are urgently required to address these challenges associated with high-pressure storage methods.

Liquid hydrogen storage avoids the problems associated with high-pressure hydrogen storage, but liquefied hydrogen requires higher energy, which results in higher costs. Another challenge of liquid hydrogen storage is the fact of hydrogen boiling or evaporation. Research in low-temperature storage of liquid hydrogen is a critical area of study, with current efforts concentrated on enhancing tank designs and materials to overcome the complexities linked to storing hydrogen at such temperatures.

5.3. Optimizing Control Methods to Improve Power Generation Efficiency

The flexible application of hydrogen to power systems through fuel cells will be an important direction of hydrogen electrification. In the research of fuel cells, the technical issues of catalyst degradation need to be more clearly understood in order to increase their power rating, extend their continuous operating time, and reduce production costs. Various components within hydrogen-based power systems typically interconnect through power electronic converters for hydrogen production and utilization. To ensure optimal system performance, converters require attributes such as adaptable voltage ratios, efficient conversion rates, and minimal current fluctuations. Research in converters for hydrogen fuel cells should prioritize effective control mechanisms to mitigate power output disruptions caused by switching faults. Concurrently, enhancing voltage ratios is essential for system reliability, thereby bolstering fuel cell power generation efficiency.

According to the literature review and research in different periods, it can be seen that the application of hydrogen energy in the power system will mainly focus on the improvement of system efficiency, service life, and durability, as well as the coupling and coordinated control of hydrogen energy and power systems. As the whole of society pays more attention to the research of hydrogen energy, under the guidance and promotion of national and local policies, a power system containing hydrogen energy will be developed in the long run.

Author Contributions: Conceptualization, S.D., Q.Y. and W.D.; methodology, S.D.; formal analysis, P.S. and Q.Y.; resources, S.D., P.S. and Q.Y.; writing—original draft preparation, S.D. and W.D.; writing—review and editing, S.D., P.S. and W.D.; visualization, P.S. and W.D.; supervision, W.D. and Q.Y.; project administration, W.D.; funding acquisition, W.D. All authors have read and agreed to the published version of the manuscript.

Funding: Guangdong Basic and Applied Basic Research Foundation: 2022A1515110794; Guangzhou Basic and Applied Basic Research Foundation: 2023A04J0973.

Data Availability Statement: The original data presented in the study can be found from the cited publications.

Conflicts of Interest: The authors declare no conflict of interest.

References

1. Yang, Y.; Li, Z.; Mandapaka, P.V.; Lo, E.Y. Risk-averse restoration of coupled power and water systems with small pumped-hydro storage and stochastic rooftop renewables. *Appl. Energy* **2023**, *339*, 120953. [CrossRef]
2. Saidi, K.; Omri, A. The impact of renewable energy on carbon emissions and economic growth in 15 major renewable energy-consuming countries. *Environ. Res.* **2020**, *186*, 109567. [CrossRef] [PubMed]
3. Zheng, H.; Song, M.; Shen, Z. The evolution of renewable energy and its impact on carbon reduction in China. *Energy* **2021**, *237*, 121639. [CrossRef]
4. Sun, Y.; Li, H.; Andlib, Z.; Genie, M.G. How do renewable energy and urbanization cause carbon emissions? Evidence from advanced panel estimation techniques. *Renew. Energy* **2022**, *185*, 996–1005. [CrossRef]
5. Behabtu, H.A.; Messagie, M.; Coosemans, T.; Berecibar, M.; Anlay Fante, K.; Kebede, A.A.; Mierlo, J. V A review of energy storage technologies' application potentials in renewable energy sources grid integration. *Sustainability* **2020**, *12*, 10511. [CrossRef]
6. Kebede, A.A.; Kalogiannis, T.; Van Mierlo, J.; Berecibar, M. A comprehensive review of stationary energy storage devices for large scale renewable energy sources grid integration. *Renew. Sustain. Energy Rev.* **2022**, *159*, 112213. [CrossRef]
7. Le, T.T.; Sharma, P.; Bora, B.J.; Tran, V.D.; Truong, T.H.; Le, H.C.; Nguyen, P.Q.P. Fueling the future: A comprehensive review of hydrogen energy systems and their challenges. *Int. J. Hydrogen Energy* **2024**, *54*, 791–816. [CrossRef]
8. Nicoletti, G.; Arcuri, N.; Nicoletti, G.; Bruno, R. A technical and environmental comparison between hydrogen and some fossil fuels. *Energy Convers. Manag.* **2015**, *89*, 205–213. [CrossRef]
9. Capurso, T.; Stefanizzi, M.; Torresi, M.; Camporeale, S.M. Perspective of the role of hydrogen in the 21st century energy transition. *Energy Convers. Manag.* **2022**, *251*, 114898. [CrossRef]
10. Chai, S.; Zhang, G.; Li, G.; Zhang, Y. Industrial hydrogen production technology and development status in China: A review. *Clean. Technol. Environ. Policy* **2021**, *23*, 1931–1946. [CrossRef]
11. Younas, M.; Shafique, S.; Hafeez, A.; Javed, F.; Rehman, F. An overview of hydrogen production: Current status, potential, and challenges. *Fuel* **2022**, *316*, 123317. [CrossRef]
12. Li, J.; Cheng, W. Comparative life cycle energy consumption, carbon emissions and economic costs of hydrogen production from coke oven gas and coal gasification. *Int. J. Hydrogen Energy* **2020**, *45*, 27979–27993. [CrossRef]
13. Yang, K.; Gu, Z.; Long, Y.; Lin, S.; Lu, C.; Zhu, X.; Wang, H.; Li, K. Hydrogen production via chemical looping reforming of coke oven gas. *Green Energy Environ.* **2021**, *6*, 678–692. [CrossRef]
14. Chi, J.; Yu, H. Water electrolysis based on renewable energy for hydrogen production. *Chin. J. Catal.* **2018**, *39*, 390–394. [CrossRef]
15. Rashid, M.D.; Al Mesfer, M.K.; Naseem, H.; Danish, M. Hydrogen production by water electrolysis: A review of alkaline water electrolysis, PEM water electrolysis and high temperature water electrolysis. *Int. J. Eng. Adv. Technol.* **2015**, *4*, 2249–8958.

16. Gong, Y.; Yao, J.; Wang, P.; Li, Z.; Zhou, H.; Xu, C. Perspective of hydrogen energy and recent progress in electrocatalytic water splitting. *Chin. J. Chem. Eng.* **2022**, *43*, 282–296. [CrossRef]
17. Li, Z.; Zhang, W.; Zhang, R.; Sun, H. Development of renewable energy multi-energy complementary hydrogen energy system (A Case Study in China): A review. *Energy Explor. Exploit.* **2020**, *38*, 2099–2127. [CrossRef]
18. Hassan, Q.; Sameen, A.Z.; Salman, H.M.; Jaszczur, M. Large-scale green hydrogen production via alkaline water electrolysis using solar and wind energy. *Int. J. Hydrogen Energy* **2023**, *48*, 34299–34315. [CrossRef]
19. Mandal, M. Recent Advancement on Anion Exchange Membranes for Fuel Cell and Water Electrolysis. *ChemElectroChem* **2021**, *8*, 36–45. [CrossRef]
20. Anwar, S.; Khan, F.; Zhang, Y.; Djire, A. Recent developments in electrocatalysts for hydrogen production through water electrolysis. *Int. J. Hydrogen Energy* **2021**, *46*, 32284–32317. [CrossRef]
21. González, G.M.C.; Toharias, B.; Iranzo, A.; Suárez, C.; Rosa, F. Voltage Distribution Analysis and Non-uniformity Assessment in a 100 cm^2 PEM Fuel Cell Stack. *Energy* **2023**, *282*, 128781. [CrossRef]
22. Sazali, N.; Wan Salleh, W.N.; Jamaludin, A.S.; Mhd Razali, M.N. New Perspectives on Fuel Cell Technology: A Brief Review. *Membranes* **2020**, *10*, 99. [CrossRef] [PubMed]
23. Yodwong, B.; Guilbert, D.; Phattanasak, M.; Kaewmanee, W.; Hinaje, M.; Vitale, G. Faraday's Efficiency Modeling of a Proton Exchange Membrane Electrolyzer Based on Experimental Data. *Energies* **2020**, *13*, 4792. [CrossRef]
24. Xu, J.; Zhang, C.; Wan, Z.; Chen, X.; Chan, S.H.; Tu, Z. Progress and Perspectives of Integrated Thermal Management Systems in PEM Fuel Cell Vehicles: A Review. *Renew. Sustain. Energy Rev.* **2022**, *155*, 111908. [CrossRef]
25. Pareek, A.; Dom, R.; Gupta, J.; Chandran, J.; Adepu, V.; Borse, P.H. Insights into renewable hydrogen energy: Recent advances and prospects. *Mater. Sci. Energy Technol.* **2020**, *3*, 319–327. [CrossRef]
26. Nasser, M.; Megahed, T.F.; Ookawara, S.; Hassan, H. A review of water electrolysis–based systems for hydrogen production using hybrid/solar/wind energy systems. *Environ. Sci. Pollut. Res.* **2022**, *29*, 86994–87018. [CrossRef]
27. Gutiérrez-Martín, F.; Amodio, L.; Pagano, M. Hydrogen production by water electrolysis and off-grid solar PV. *Int. J. Hydrogen Energy* **2021**, *46*, 29038–29048. [CrossRef]
28. Nemiwal, M.; Zhang, T.C.; Kumar, D. Graphene-based electrocatalysts: Hydrogen evolution reactions and overall water splitting. *Int. J. Hydrogen Energy* **2021**, *46*, 21401–21418. [CrossRef]
29. Das, C.K.; Bass, O.; Kothapalli, G.; Mahmoud, T.S.; Habibi, D. Overview of energy storage systems in distribution networks: Placement, sizing, operation, and power quality. *Renew. Sustain. Energy Rev.* **2018**, *91*, 1205–1230. [CrossRef]
30. Palizban, O.; Kauhaniemi, K. Energy storage systems in modern grids—Matrix of technologies and applications. *J. Energy Storage* **2016**, *6*, 248–259. [CrossRef]
31. Yao, L.; Yang, B.; Cui, H.; Zhuang, J.; Ye, J.; Xue, J. Challenges and progresses of energy storage technology and its application in power systems. *J. Mod. Power Syst. Clean Energy* **2016**, *4*, 519–528. [CrossRef]
32. Sabihuddin, S.; Kiprakis, A.E.; Mueller, M. A numerical and graphical review of energy storage technologies. *Energies* **2015**, *8*, 172–216. [CrossRef]
33. Nadeem, F.; Hussain, S.M.S.; Tiwari, P.K.; Goswami, A.K.; Ustun, T.S. Comparative review of energy storage systems, their roles, and impacts on future power systems. *IEEE Access* **2019**, *7*, 4555–4585. [CrossRef]
34. Maisanam, A.K.S.; Biswas, A.; Sharma, K.K. An innovative framework for electrical energy storage system selection for remote area electrification with renewable energy system: Case of a remote village in India. *J. Renew. Sustain. Energy* **2020**, *12*, 024101. [CrossRef]
35. Koohi-Fayegh, S.; Rosen, M.A. A review of energy storage types, applications and recent developments. *J. Energy Storage* **2020**, *27*, 101047. [CrossRef]
36. Alves, M.P.; Gul, W.; Cimini Junior, C.A.; Ha, S.K. A review on industrial perspectives and challenges on material, manufacturing, design and development of compressed hydrogen storage tanks for the transportation sector. *Energies* **2022**, *15*, 5152. [CrossRef]
37. Hassan, I.A.; Ramadan, H.S.; Saleh, M.A.; Hissel, D. Hydrogen storage technologies for stationary and mobile applications: Review, analysis and perspectives. *Renew. Sust. Energ Rev.* **2021**, *149*, 111311. [CrossRef]
38. Rahman, M.M.; Oni, A.O.; Gemechu, E.; Kumar, A. Assessment of energy storage technologies: A review. *Energy Convers. Manag.* **2020**, *223*, 113295. [CrossRef]
39. Yan, Y.; Zhang, J.; Li, G.; Zhou, W.; Ni, Z. Review on linerless type V cryo-compressed hydrogen storage vessels: Resin toughening and hydrogen-barrier properties control. *Renew. Sustain. Energy Rev.* **2024**, *189*, 114009. [CrossRef]
40. Noh, H.; Kang, K.; Seo, Y. Environmental and Energy Efficiency Assessments of Offshore Hydrogen Supply Chains Utilizing Compressed Gaseous Hydrogen, Liquefied Hydrogen, Liquid Organic Hydrogen Carriers and Ammonia. *Int. J. Hydrogen Energy* **2023**, *48*, 7515–7532. [CrossRef]
41. Chang, J.; Huang, Q.; Gao, Y.; Zheng, Z. Pathways of Developing High-Energy-Density Flexible Lithium Batteries. *Adv. Mater.* **2021**, *33*, 2004419. [CrossRef]
42. Cigolotti, V.; Genovese, M.; Fragiacomo, P. Comprehensive Review on Fuel Cell Technology for Stationary Applications as Sustainable and Efficient Poly-Generation Energy Systems. *Energies* **2021**, *14*, 4963. [CrossRef]
43. Cheng, X.B.; Liu, H.; Yuan, H.; Peng, H.J.; Tang, C.; Huang, J.Q.; Zhang, Q. A Perspective on Sustainable Energy Materials for Lithium Batteries. *SusMat* **2021**, *1*, 38–50. [CrossRef]

44. Lai, X.; Huang, Y.; Deng, C.; Gu, H.; Han, X.; Zheng, Y.; Ouyang, M. Sorting, Regrouping, and Echelon Utilization of the Large-Scale Retired Lithium Batteries: A Critical Review. *Renew. Sustain. Energy Rev.* **2021**, *146*, 111162. [CrossRef]
45. Jiao, K.; Xuan, J.; Du, Q.; Bao, Z.; Xie, B.; Wang, B.; Zhao, Y.; Fan, L.; Wang, H.; Hou, Z.; et al. Designing the Next Generation of Proton-Exchange Membrane Fuel Cells. *Nature* **2021**, *595*, 361–369. [CrossRef] [PubMed]
46. Luo, Y.; Wu, Y.; Li, B.; Mo, T.; Li, Y.; Feng, S.P.; Qu, J.; Chu, P.K. Development and Application of Fuel Cells in the Automobile Industry. *J. Energy Storage* **2021**, *42*, 103124. [CrossRef]
47. Zhao, X.; Yan, Y.; Zhang, J.; Zhang, F.; Wang, Z.; Ni, Z. Analysis of multilayered carbon fiber winding of cryo-compressed hydrogen storage vessel. *Int. J. Hydrogen Energy* **2022**, *47*, 10934–10946. [CrossRef]
48. Tarkowski, R.; Uliasz-Misiak, B. Towards underground hydrogen storage: A review of barriers. *Renew. Sustain. Energy Rev.* **2022**, *162*, 112451. [CrossRef]
49. Zivar, D.; Kumar, S.; Foroozesh, J. Underground hydrogen storage: A comprehensive review. *Renew. Sustain. Energy Rev.* **2021**, *46*, 23436–23462. [CrossRef]
50. Thiyagarajan, S.R.; Emadi, H.; Hussain, A.; Patange, P.; Watson, M. A comprehensive review of the mechanisms and efficiency of underground hydrogen storage. *J. Energy Storage* **2022**, *51*, 104490. [CrossRef]
51. Kovač, A.; Paranos, M.; Marciuš, D. Hydrogen in energy transition: A review. *Int. J. Hydrogen Energy* **2021**, *46*, 10016–10035. [CrossRef]
52. Yin, L.; Ju, Y. Review on the design and optimization of hydrogen liquefaction processes. *Front. Energy* **2020**, *14*, 530–544. [CrossRef]
53. Milanese, C.; Jensen, T.R.; Hauback, B.C.; Pistidda, C.; Dornheim, M.; Yang, H.; Lombardo, L.; Zuettel, A.; Filinchuk, Y.; Ngene, P.; et al. Complex hydrides for energy storage. *Int. J. Hydrogen Energy* **2019**, *44*, 7860–7874. [CrossRef]
54. Shang, Y.; Pistidda, C.; Gizer, G.; Klassen, T.; Dornheim, M. Mg-based materials for hydrogen storage. *J. Magnes Alloy* 2021, *in press*. [CrossRef]
55. Yang, J.; Li, Y.; Tan, H. An energy-saving hydrogen liquefaction process with efficient utilization of liquefied natural gas cold energy. *Int. J. Hydrogen Energy* **2024**, *49*, 1482–1496. [CrossRef]
56. Morales-Ospino, R.; Celzard, A.; Fierro, V. Strategies to recover and minimize boil-off losses during liquid hydrogen storage. *Renew. Sustain. Energy Rev.* **2023**, *182*, 113360. [CrossRef]
57. Weckerle, C.; Dörr, M.; Linder, M.; Bürger, I. A compact thermally driven cooling system based on metal hydrides. *Energies* **2020**, *13*, 2482. [CrossRef]
58. Wang, J.; Qi, M.; Liu, S.; Zhao, D. Comprehensive safety assessment of a hydrogen liquefaction system based on an integrated system-theoretic process analysis (STPA) and best-worst method (BWM). *Int. J. Hydrogen Energy* **2024**, *66*, 479–489. [CrossRef]
59. Argyrou, M.C.; Christodoulides, P.; Wongwises, S.A. Energy storage for electricity generation and related processes: Technologies appraisal and grid scale applications. *Renew. Sustain. Energy Rev.* **2018**, *94*, 804–821. [CrossRef]
60. Ghorbani, B.; Zendehboudi, S.; Saady, N.M.C.; Duan, X.; Albayati, T.M. Strategies to improve the performance of hydrogen storage systems by liquefaction methods: A comprehensive review. *ACS Omega* **2023**, *8*, 18358–18399. [CrossRef] [PubMed]
61. Chu, C.; Wu, K.; Luo, B.; Cao, Q.; Zhang, H. Hydrogen storage by liquid organic hydrogen carriers: Catalyst, renewable carrier, and technology–a review. *Carbon Resour. Convers.* **2023**, *6*, 334–351. [CrossRef]
62. Tan, K.C.; Chua, Y.S.; He, T.; Chen, P. Strategies of thermodynamic alternation on organic hydrogen carriers for hydrogen storage application: A review. *Green Energy Resour.* **2023**, *1*, 100020. [CrossRef]
63. Pingkuo, P.L.; Xue, H. Comparative analysis on similarities and differences of hydrogen energy development in the world's top 4 largest economies: A novel framework. *Int. J. Hydrogen Energy* **2022**, *47*, 9485–9503. [CrossRef]
64. Zhai, L.; Liu, S.; Xiang, Z. Ammonia as a carbon-free hydrogen carrier for fuel cells: A perspective. *Ind. Chem. Mater.* **2023**, *1*, 332–342. [CrossRef]
65. Negro, V.; Noussan, M.; Chiaramonti, D. The potential role of ammonia for hydrogen storage and transport: A critical review of challenges and opportunities. *Energies* **2023**, *16*, 6192. [CrossRef]
66. Kojima, Y. Safety of ammonia as a hydrogen energy carrier. *Int. J. Hydrogen Energy* **2024**, *50*, 732–739. [CrossRef]
67. Ramezani, R.; Di Felice, L.; Gallucci, F. A review of chemical looping reforming technologies for hydrogen produc-tion: Recent advances and future challenges. *J. Phys. Energy* **2023**, *5*, 024010. [CrossRef]
68. Guan, D.; Wang, B.; Zhang, J.; Shi, R.; Jiao, K.; Li, L.; Wang, Y.; Xie, B.; Zhang, Q.; Yu, J.; et al. Hydrogen society: From present to future. *Energy Environ. Sci.* **2023**, *16*, 4926–4943. [CrossRef]
69. Zhang, L.; Jia, C.; Bai, F.; Wang, W.; An, S.; Zhao, K.; Li, Z.; Li, J.; Sun, H. A comprehensive review of the promising clean energy carrier: Hydrogen production, transportation, storage, and utilization (HPTSU) technologies. *Fuel* **2024**, *355*, 129455. [CrossRef]
70. Klopčič, N.; Grimmer, I.; Winkler, F.; Sartory, M.; Trattner, A. A review on metal hydride materials for hydrogen storage. *J Energy Storage* **2023**, *72*, 108456. [CrossRef]
71. Srinivasan, M.N.; Babu, M.D.; Naveena, S.S.; Malini, A.S.; Ilyos, K.; Dixit, K.K.; Husain, S.O.; Dev, B.P.V. Metal hydride hydrogen storage and compression systems for energy storage: Modeling, synthesis, and properties. In Proceedings of the 2023 International Conference for Technological Engineering and its Applications in Sustainable Development (ICTEASD), Al-Najaf, Iraq, 14–15 November 2023; IEEE: Piscataway, NJ, USA, 2023; pp. 256–261.
72. Zhang, X.; Lou, Z.; Gao, M.; Pan, H.; Liu, Y. Metal hydrides for advanced hydrogen/lithium storage and ionic conduction applications. *Acc. Mater. Res.* **2024**, *5*, 371–384. [CrossRef]

73. Dubey, S.K.; Ravi Kumar, K.; Tiwari, V.; Srivastva, U. Impacts, barriers, and future prospective of metal hydride-based thermochemical energy storage system for high-temperature applications: A comprehensive review. *Energy Technol.* **2024**, *12*, 2300768. [CrossRef]
74. Franke, F.; Kazula, S.; Enghardt, L. Review and evaluation of metal-hydride-based hydrogen sensors as safety devices for future sustainable aviation. *Proc. J. Phys. Conf. Ser.* **2023**, *2454*, 012001. [CrossRef]
75. Nivedhitha, K.S.; Beena, T.; Banapurmath, N.R.; Umarfarooq, M.A.; Ramasamy, V.; Soudagar, M.E.M.; Ağbulut, Ü. Advances in hydrogen storage with metal hydrides: Mechanisms, materials, and challenges. *Int. J. Hydrogen Energy* **2024**, *61*, 1259–1273. [CrossRef]
76. Evans, A.; Strezov, V.; Evans, T.J. Assessment of utility energy storage options for increased renewable energy penetration Sustain. *Energy Rev.* **2012**, *16*, 4141–4147.
77. Gao, J.; Zhou, S.; Lu, Y.; Shen, W. Simulation of a Novel Integrated Multi-Stack Fuel Cell System Based on a Double-Layer Multi-Objective Optimal Allocation Approach. *Appl. Sci.* **2024**, *14*, 2961. [CrossRef]
78. Xie, Z.; Zhou, S.; Gao, J.; Zhang, G.; Shen, W. Structural Design, Matching, and Analysis of Air Supply Devices for Multi-Stack Fuel Cell Systems. *Energy Technol.* **2023**, *11*, 2201331. [CrossRef]
79. Ghaderi, R.; Kandidayeni, M.; Boulon, L.; Trovão, J.P. Q-learning based energy management strategy for a hybrid multi-stack fuel cell system considering degradation. *Energy Convers. Manag.* **2023**, *293*, 117524. [CrossRef]
80. Ding, D.; Wu, X.Y. Hydrogen Fuel Cell Electric Trains: Technologies, Current Status, and Future. *Appl. Energy Combust. Sci.* **2024**, *17*, 100255. [CrossRef]
81. Li, H.; Guo, H.; Yousefi, N. A Hybrid Fuel Cell/Battery Vehicle by Considering Economy Considerations Optimized by Converged Barnacles Mating Optimizer (CBMO) Algorithm. *Energy Rep.* **2020**, *6*, 2441–2449. [CrossRef]
82. Hoskins, A.L.; Millican, S.L.; Czernik, C.E.; Alshankiti, I.; Netter, J.C.; Wendelin, T.J.; Musgrave, C.B.; Weimer, A.W. Continuous On-Sun Solar Thermochemical Hydrogen Production via an Isothermal Redox Cycle. *Appl. Energy* **2019**, *249*, 368–376. [CrossRef]
83. Jacobsen, A.; Godvik, M. Influence of Wakes and Atmospheric Stability on the Floater Responses of the Hywind Scotland Wind Turbines. *Wind Energy* **2021**, *24*, 149–161. [CrossRef]
84. Sharpe, O.; Raman, K. The Australian Hydrogen Centre–Feasibility Studies for Achieving 10 and 100% Renewable Hydrogen in South Australia and Victoria. *Aust. Energy Prod. J.* **2024**, *64*, S191–S196. [CrossRef]
85. Biabani, H.; Aminlou, A.; Hayati, M.M.; Majidi-Gharehnaz, H.; Abapour, M. Green Hydrogen Research and Development Projects in the European Union. In *Green Hydrogen in Power Systems*; Springer International Publishing: Berlin/Heidelberg, Germany, 2024; pp. 301–320.
86. Durakovic, G.; del Granado, P.C.; Tomasgard, A. Powering Europe with North Sea Offshore Wind: The Impact of Hydrogen Investments on Grid Infrastructure and Power Prices. *Energy* **2023**, *263*, 125654. [CrossRef]
87. Ishaq, H.; Dincer, I.; Crawford, C. A review on hydrogen production and utilization: Challenges and opportunities. *Int. J. Hydrogen Energy* **2022**, *47*, 26238–26264. [CrossRef]
88. Ragab, A.; Marei, M.I.; Mokhtar, M. Comprehensive Study of Fuel Cell Hybrid Electric Vehicles: Classification, Topologies, and Control System Comparisons. *Appl. Sci.* **2023**, *13*, 13057. [CrossRef]
89. Ma, Y.; Li, C.; Wang, S. Multi-Objective Energy Management Strategy for Fuel Cell Hybrid Electric Vehicle Based on Stochastic Model Predictive Control. *ISA Trans.* **2022**, *131*, 178–196. [CrossRef] [PubMed]
90. Li, X.; Shang, Z.; Peng, F.; Li, L.; Zhao, Y.; Liu, Z. Increment-oriented online power distribution strategy for multi-stack proton exchange membrane fuel cell systems aimed at collaborative performance enhancement. *J. Power Sources* **2021**, *512*, 230512–230528. [CrossRef]
91. Ameli, H.; Strbac, G.; Pudjianto, D.; Ameli, M.T. A Review of the Role of Hydrogen in the Heat Decarbonization of Future Energy Systems: Insights and Perspectives. *Energies* **2024**, *17*, 1688. [CrossRef]
92. Shi, W.; Huangfu, Y.; Xu, L.; Pang, S. Online energy management strategy considering fuel cell fault for multi-stack fuel cell hybrid vehicle based on multi-agent reinforcement learning. *Appl. Energy* **2022**, *328*, 120234. [CrossRef]
93. İskenderoğlu, F.C.; Baltacioğlu, M.F.; Demir, M.H.; Baldinelli, A.; Barelli, L.; Bidini, G. Comparison of support vector regression and random forest algorithms for estimating the SOFC output voltage by considering hydrogen flow rates. *Int. J. Hydrogen Energy* **2020**, *45*, 35023–35038. [CrossRef]
94. Ban, M.; Yu, J.; Shahidehpour, M.; Yao, Y. Integration of power-to-hydrogen in day-ahead security-constrained unit commitment with high wind penetration. *J. Mod. Power Syst. Clean Energy* **2017**, *5*, 337–349. [CrossRef]
95. Boretti, A. A high-efficiency internal combustion engine using oxygen and hydrogen. *Int. J. Hydrogen Energy* **2024**, *50*, 847–856. [CrossRef]
96. Zhang, S.W.; Sun, B.G.; Lin, S.L.; Li, Q.; Wu, X.; Hu, T.; Bao, B.-Z.; Wang, X.; Luo, Q.-H. Energy and exergy analysis for a turbocharged direct-injection hydrogen engine to achieve efficient and high-economy performances. *Int. J. Hydrogen Energy* **2024**, *54*, 601–612. [CrossRef]
97. Yu, Z.; Han, J.; Cao, X. Investigation on performance of an integrated solid oxide fuel cell and absorption chiller tri-generation system. *Int. J. Hydrogen Energy* **2011**, *36*, 12561–12573. [CrossRef]
98. Da Silva, D.C.; Kefsi, L.; Sciarretta, A. Closed-Form Expression to Estimate the Hydrogen Consumption of a Fuel Cell Hybrid Electric Vehicle. *IEEE Trans. Veh. Technol.* **2024**, *73*, 4717–4728. [CrossRef]

99. Hydrogen Council Path to Hydrogen Competitiveness: A Cost Perspective. 2020. Available online: https://hydrogencouncil.com/en/path-to-hydrogen-competitiveness-a-cost-perspective// (accessed on 20 January 2020).
100. Gu, Z.; Pan, G.; Gu, W.; Qiu, H.; Lu, S. Robust Optimization of Scale and Revenue for Integrated Power-to-Hydrogen Systems within Energy, Ancillary Services, and Hydrogen Markets. *IEEE Trans. Power Syst.* **2023**, *39*, 5008–5023. [CrossRef]
101. Yan, C.; Zou, Y.; Wu, Z.; Maleki, A. Effect of Various Design Configurations and Operating Conditions for Optimization of a Wind/Solar/Hydrogen/Fuel Cell Hybrid Microgrid System by a Bio-Inspired Algorithm. *Int. J. Hydrogen Energy* **2024**, *60*, 378–391. [CrossRef]
102. Deng, W.; Zhang, Y.; Tang, Y.; Li, Q.; Yi, Y. A Neural Network-Based Adaptive Power-Sharing Strategy for Hybrid Frame Inverters in a Microgrid. *Front. Energy Res.* **2023**, *10*, 1082948. [CrossRef]
103. Fu, Z.; Lu, L.; Zhang, C.; Xu, Q.; Zhang, X.; Gao, Z.; Li, J. Fuel Cell and Hydrogen in Maritime Application: A Review on Aspects of Technology, Cost, and Regulations. *Sustain. Energy Technol. Assess.* **2023**, *57*, 103181. [CrossRef]
104. Yang, X.; Sun, J.; Meng, X.; Sun, S.; Shao, Z. Cold Start Degradation of Proton Exchange Membrane Fuel Cell: Dynamic and Mechanism. *Chem. Eng. J.* **2023**, *455*, 140823. [CrossRef]
105. Teoh, Y.H.; How, H.G.; Le, T.D.; Nguyen, H.T.; Loo, D.L.; Rashid, T.; Sher, F. A Review on Production and Implementation of Hydrogen as a Green Fuel in Internal Combustion Engines. *Fuel* **2023**, *333*, 126525. [CrossRef]
106. Park, Y.K.; Kim, B.S. Catalytic Removal of Nitrogen Oxides (NO, NO_2, N_2O) from Ammonia-Fueled Combustion Exhaust: A Review of Applicable Technologies. *Chem. Eng. J.* **2023**, *461*, 141958. [CrossRef]
107. Albatayneh, A.; Juaidi, A.; Jaradat, M.; Manzano-Agugliaro, F. Future of Electric and Hydrogen Cars and Trucks: An Overview. *Energies* **2023**, *16*, 3230. [CrossRef]
108. Dall'Armi, C.; Pivetta, D.; Taccani, R. Hybrid PEM Fuel Cell Power Plants Fuelled by Hydrogen for Improving Sustainability in Shipping: State of the Art and Review on Active Projects. *Electronics* **2023**, *16*, 2022. [CrossRef]
109. Klatzer, T.; Bachhiesl, U.; Wogrin, S. State-of-the-art expansion planning of integrated power, natural gas, and hydrogen systems. *Int. J. Hydrogen Energy* **2022**, *47*, 20585–20603. [CrossRef]
110. Gonzalez-Romero, I.C.; Wogrin, S.; Gómez, T. Review on generation and transmission expansion co-planning models under a market environment. *IET Gener. Transm. Dis.* **2020**, *14*, 931–944. [CrossRef]
111. Zhang, S.; Zhang, N.; Dai, H.; Liu, L.; Zhou, Z.; Shi, Q.; Lu, J. Comparison of Different Coupling Modes between the Power System and the Hydrogen System Based on a Power–Hydrogen Coordinated Planning Optimization Model. *Energies* **2023**, *16*, 5374. [CrossRef]
112. Khaleel, M.M.; Adzman, M.R.; Zali, S.M. An Integrated Hydrogen Fuel Cell to Distribution Network System: Challenges and Opportunities for D-STATCOM. *Energies* **2021**, *14*, 7073. [CrossRef]
113. Qiu, Y.; Li, Q.; Ai, Y.; Chen, W.; Benbouzid, M.; Liu, S.; Gao, F. Two-Stage Distributionally Robust Optimization-Based Coordinated Scheduling of Integrated Energy System with Electricity-Hydrogen Hybrid Energy Storage. *Prot. Control Mod. Power Syst.* **2023**, *8*, 542–555. [CrossRef]
114. Yan, Y. Comparative Analysis of the Economics of Hydrogen Storage and Transportation. Master's Thesis, Huazhong University of Science Technology, Wuhan, China, 2021.
115. Salama, H.S.; Magdy, G.; Bakeer, A.; Vokony, I. Adaptive Coordination Control Strategy of Renewable Energy Sources, Hydrogen Production Unit, and Fuel Cell for Frequency Regulation of a Hybrid Distributed Power System. *Prot. Control Mod. Power Syst.* **2022**, *7*, 472–489. [CrossRef]
116. Marquez, R.A.; Espinosa, M.; Kalokowski, E.; Son, Y.J.; Kawashima, K.; Le, T.V.; Chukwuneke, C.E.; Mullins, C.B. A Guide to Electrocatalyst Stability Using Lab-Scale Alkaline Water Electrolyzers. *ACS Energy Lett.* **2024**, *9*, 547–555. [CrossRef]
117. Rocha, C.; Knöri, T.; Ribeirinha, P.; Gazdzicki, P. A review on flow field design for proton exchange membrane fuel cells: Challenges to increase the active area for MW applications. *Renew. Sustain. Energy Rev.* **2024**, *192*, 114198. [CrossRef]
118. Caldera, U.; Bogdanov, D.; Afanasyeva, S.; Breyer, C. Role of seawater desalination in the management of an integrated water and 100% renewable energy based power sector in Saudi Arabia. *Water* **2017**, *10*, 3. [CrossRef]

electronics

MDPI

Article

A Cooperative Operation Strategy for Multi-Energy Systems Based on the Power Dispatch Meta-Universe Platform

Jinbo Liu [1,2], Lijuan Duan [1,*], Jian Chen [3], Jingan Shang [3], Bin Wang [1] and Zhaoguang Pan [1]

1 Department of Electrical Engineering, Tsinghua University, Beijing 100084, China; liu-jinbo@sgcc.com.cn (J.L.); wb1984@tsinghua.edu.cn (B.W.); panzhaoguang@tsinghua.edu.cn (Z.P.)
2 National Power Dispatching and Control Center, State Grid Corporation of China, Beijing 100031, China
3 State Grid Tianjin Electric Power Company, Tianjin 300310, China; jian.chen2@tj.sgcc.com.cn (J.C.); jingan.shang@tj.sgcc.com.cn (J.S.)
* Correspondence: duanlijuan@tsinghua.edu.cn

Abstract: To meet the challenges of renewable energy consumption and improve the efficiency of energy systems, we propose an intelligent distributed energy dispatch strategy for multi-energy systems based on Nash bargaining by utilizing the power dispatch meta-universe platform. First, the operational framework of the multi-energy system, including wind park (WP), photovoltaic power plant (PVPP), and energy storage (ES), is described. Using the power dispatch meta-universe platform, the models of WP, PVPP, and ES are constructed and analyzed. Then, a Nash bargaining model of the multi-energy system is built and transformed into a coalition profit maximization problem, which is solved using the alternating direction multiplier method (ADMM). Finally, the effectiveness of the proposed strategy is verified. The results show that the strategy greatly improves the consumption of renewable energy sources and the profit of the overall system.

Keywords: alternating direction multiplier method; energy storage; multi-energy system; Nash bargaining; renewable energy; power dispatch meta-universe

check for updates

Citation: Liu, J.; Duan, L.; Chen, J.; Shang, J.; Wang, B.; Pan, Z. A Cooperative Operation Strategy for Multi-Energy Systems Based on the Power Dispatch Meta-Universe Platform. *Electronics* **2024**, *13*, 3015. https://doi.org/10.3390/electronics13153015

Academic Editor: Enrique Rosales-Asensio

Received: 15 May 2024
Revised: 5 July 2024
Accepted: 6 July 2024
Published: 31 July 2024

1. Introduction

With the proposal of the "dual carbon" target, the traditional power supply system is gradually shifting towards a new type of power supply system with an increasing proportion of renewable energy [1]. Renewable energy capacity additions have risen rapidly in recent years [2]. However, renewable energy such as wind and photovoltaic (PV) power have shortcomings including randomness, intermittency, and volatility, resulting in low power quality. The large-scale grid integration of renewable energy can lead to voltage instability, further complicating grid control and dispatch operations. Thus, it has posed a huge challenge to the stable operation of the power system [3].

How to deal with uncertainty and increase penetration of renewable energy has become a research priority [4]. Reference [5] develops a stochastic optimization model for renewable energy sources to deal with the uncertainty of renewable energy sources. Reference [6] proposes a robust optimization model for microgrids, considering the uncertainty of renewable distributed energy sources (PV, wind, etc.), which improves the economics and robustness of microgrid operation. Reference [7] illustrates the ability of demand response to balance the volatility of renewable energy generation, thereby promoting higher penetration of renewable energy in the power system. However, stochastic optimization methods usually require accurate knowledge of the probability distribution of uncertainty, which may lead to unstable or imprecise optimization results if the probability distribution is inaccurate or difficult to obtain. Robust optimization tends to be more conservative, leading to situations where the system fails to maximize its potential profit or efficiency. Demand response may affect users' living and working habits, reducing their comfort and convenience. Energy storage (ES) can cope with the uncertainty of renewable energy by using a medium to store electrical energy and releasing it for power generation when there is a demand for its utilization [8], which is one of the most important ways to increase the penetration of renewable energy sources [9]. Reference [10] overviews various energy storage technologies for handling fluctuations and uncertainties. Reference [11] puts forward the idea that the integration of ES emerges as a viable solution for supporting renewable energy sources integration. Reference [12] investigates the value of seasonal energy storage technologies for wind and PV power integration.

Although the above literature suggests that ES can cope with uncertainty and increase the penetration of renewable energy, it does not fully exploit the potential for synergistic operation of renewable energy and ES. Multi-energy systems can integrate a variety of energy resources and improve the diversity and stability of the energy supply [13]. It can optimize the efficiency of energy use by coordinating the complementarity and integration of different energy systems. Research on multi-energy systems is of great significance in addressing energy challenges and is one of the important strategic directions for promoting sustainable development in the energy sector [14]. Based on above, establishing a multi-energy system with renewable energy and ES can be an effective way to explore the potential for synergistic operation of renewable energy and ES. However, the traditional dispatching command space cannot break through the geospatial limitations. Relying on the traditional methods of communication, modelling, simulation, and analysis makes it difficult to adapt to the needs of power grid operation [15]. Thus, the strategy to balance the dispatch of multi-energy systems and improve overall operational profit is an urgent issue to be researched.

Regarding the dispatch strategy, meta-universe, as a new process of the digital revolution, is a highly interactive and ultra-temporal digital ecosystem that integrates multiple new technologies [16]. Through holographic construction, holographic simulation, and fusion of virtual world and real world, meta-universe technology is able to realize the complete mapping and real-time interactions of a physical power grid in virtual space. It can respond to the challenges of operational complexity and security risks brought by the construction of renewable power systems in an all-round way. The power dispatch meta-universe makes use of big data, modern communication, artificial intelligence, internet of things, and other technologies to mobilize widely distributed source network, load, and storage resources and realize overall coordinated control of the system. It is capable of solving the problem of grid stability control after the access of renewable energy sources with large random fluctuations and a high proportion of power electronic components. By constructing digital doppelgangers and avatars in the digital dispatch command space, it is able to break through the limitations of geographic space, strengthen the degree of synergy between all levels of dispatch, and greatly enhance the synergy of fault disposal and other aspects. Therefore, the distributed cooperative dispatch of energy sources in the system based on the power dispatch meta-universe platform can help to solve the problems faced by power dispatch in power systems with a high proportion of renewable energy sources.

Regarding improvement of overall operational profit, in recent years, many scholars have carried out research on game strategies for energy systems. In game theory, non-cooperative and cooperative games are effective methods for dealing with complex interest relationships among participants. Non-cooperative games emphasize autonomous decision-making among individuals, namely individual rationality. They mainly focus on the competitive relationships among different participants and cannot achieve social optimality. Reference [17] considers the user satisfaction based on the traditional game models and establishes a Stackelberg leader–follower game model with energy stations as the decision-maker and the user as responders. The model is solved using the distributed algorithm of the beetle antenna. Reference [18] constructs a multi-leader multi-follower Stackelberg game model and studies the interaction problem between multiple distributed energy stations and multiple energy users. Reference [19] establishes a real-time supply–demand interaction model for power systems based on the Stackelberg game. Reference [20] establishes a multi-agent game decision-making model based on evolutionary games, achieving collaborative optimization of multi-agent operations. However, the above research focuses on the competitive relationships among different participants, but the potential for cooperation among participants is insufficiently considered [21], which often leads to nonsocial optimality.

Cooperative games can balance the unity of individual rationality and overall rationality, and they typically achieve global or Pareto optimality. Participants can achieve better results by working together than by acting alone, as they can coordinate their actions with each other to optimize the results. Stable solutions can be reached through negotiation or agreement, which can be reliable long-term strategies because participants tend to stick to the agreements they reach. Commonly used methods of cooperative games for handling interest relationships include Shapley value [22], Nucleolus [23], and Nash bargaining theory [24]. The Shapley value and Nucleolus method have higher computational costs in situations with many participants, while the computational efficiency of Nash bargaining does not significantly change with the number of participants. To explore the potential for cooperation among participants with computational costs considered, this article considers using Nash bargaining, which has been widely used in the field of energy systems. Reference [25] proposes Nash bargaining-based collaborative energy management for regional integrated energy systems. Reference [26] constructs a Nash bargaining model for energy sharing between micro-energy grids and energy storage. Reference [27] proposes a general Nash bargaining-based framework to depict energy trading among autonomous prosumers. Reference [28] proposes a peer-to-peer energy trading model combining shared energy storage based on asymmetric Nash bargaining theory. Reference [29] proposes a distributed cooperative operation strategy for multi-agent energy systems based on Nash bargaining. Reference [30] proposes a cooperative operative model for the wind–solar–hydrogen multi-agent energy system based on Nash bargaining theory. Reference [31] uses Nash bargaining to construct a low-carbon cooperative game model for a multi-electricity–gas interconnection system. Reference [32] uses the Nash bargaining model to simulate the game behavior of various entities. Reference [33] also uses the Nash bargaining method to describe the economic interactions between community energy managers and PV consumers in order to improve individual and social benefits. The above literature indicates that Nash bargaining can effectively solve the profit improvement problem of multi-energy systems, and this method can protect the privacy of participants by applying the alternating direction multiplier method (ADMM) [34].

Based on the above analysis, this paper proposes a Nash bargaining-based intelligent distributed energy dispatch strategy for multi-energy systems using the power dispatch meta-universe platform. The main contributions are as follows:

(1) Based on the power meta-universe platform, wind park (WP), PV power plant (PVPP), and ES models are constructed, and an operation model for multi-energy systems is established, where ES can effectively suppress the randomness and volatility of wind and PV power generation so that the consumption of wind and PV power can be promoted.

(2) An innovative incentive energy dispatch strategy based on Nash bargaining is proposed for the multi-energy system with WP, PVPP, and ES. The mechanism can motivate WP, PVPP, and ES to cooperate for improving the alliance profit.

(3) A distributed algorithm based on ADMM is designed to solve the problem of maximizing alliance profit, while the privacy of WP, PVPP, and ES is effectively preserved.

The remainder of this paper is organized as follows: Section 2 gives the problem description of the cooperative operation model of the multi-energy system based on the power dispatch meta-universe platform. Section 3 gives the mathematic formulation, including the models of WP, PVPP, and ES. Section 4 gives the Nash bargaining model and the distributed solving method based on ADMM. Section 5 performs case studies. Finally, Section 6 draws the conclusions.

2. Problem Description

A power dispatch meta-universe platform is built to address the new demands of power dispatch operations under the power system with a high proportion renewable energy, enhancing system reliability and stability. It has a virtual simulation function that can simulate different dispatching scenarios. Therefore, we built a typical multi-energy system with WP, PVPP, and ES based on this platform in order to achieve efficient energy dispatching, as shown in Figure 1. In the meta-universe platform, the WP, PVPP, and ES are all connected to the power grid and belong to different stakeholders. To promote the consumption of renewable energy, direct transactions of energy are allowed between the distributed generations and power consumers through the grid, while the grid will charge the network fee to recover the maintenance fee of the grid infrastructure [30].

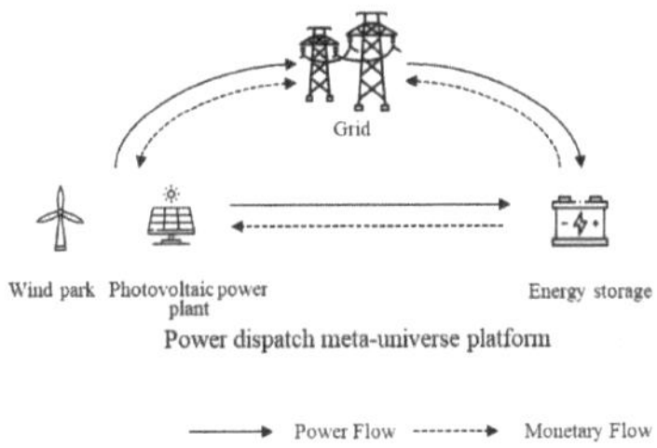

Figure 1. The cooperative operation model of the multi-energy system based on the power dispatch meta-universe platform.

In the cooperative operation model, WP, PVPP, and ES are considered as an alliance. ES can purchase power from WP and PVPP directly through negotiation, and WP and PVPP pay the network fees to the power grid. Through the power dispatch meta-universe platform, we can simulate the operation of WP, PVPP, and ES, which is conducive to enhancing the alliance profit.

3. Mathematic Formulation

3.1. Model of WP

WP generates profits U_{WP} through transactions with ES and the power grid:

$$U_{\mathrm{WP}} = U_{\mathrm{WP2ES}} + U_{\mathrm{WP2PG}} \tag{1}$$

$$U_{\mathrm{WP2ES}} = \sum_{t=1}^{T} p^t_{\mathrm{WP2ES}} P^t_{\mathrm{WP2ES}} \tag{2}$$

$$U_{\mathrm{WP2PG}} = \sum_{t=1}^{T} p^t_{\mathrm{WP2PG}} P^t_{\mathrm{WP2PG}} \tag{3}$$

where U_{WP2ES} and U_{WP2PG} are the profits made from the transaction with ES and the power grid, respectively; p^t_{WP2ES} is the trading price with ES at time t; p^t_{WP2PG} is the wind power feed-in tariff at time t; and P^t_{WP2ES} and P^t_{WP2PG} are the electricity sold to ES and the power grid, respectively [30].

The costs associated with WP C_{WP} include the maintenance fee, C_{WPM} and the network fee, C_{WPN}:

$$C_{\mathrm{WP}} = C_{\mathrm{WPM}} + C_{\mathrm{WPN}} \tag{4}$$

$$C_{\mathrm{WPM}} = \sum_{t=1}^{T} \omega_{\mathrm{WP}} P^t_{\mathrm{WP}} \tag{5}$$

$$C_{\text{WPN}} = \sum_{t=1}^{T} [\alpha_{\text{WP}} (P_{\text{WP2ES}}^t)^2 + \beta_{\text{WP}} P_{\text{WP2ES}}^t] \quad (6)$$

where ω_{WP} is the coefficient of the WP maintenance fee; P_{WP}^t is the electricity produced by WP at time t; and α_{WP} and β_{WP} are the network fee coefficients of WP.

The electricity produced by WP should satisfy the power balance constraint and should be less than the maximum electricity generation of WP, as follows:

$$P_{\text{WP}}^t = P_{\text{WP2ES}}^t + P_{\text{WP2PG}}^t \quad (7)$$

$$0 \le P_{\text{WP}}^t \le P_{\text{WP}}^{\max} \quad (8)$$

where $P_{\text{WP}}^{\max}$ is the maximum electricity generation of WP.

The objective function for WP, aimed at profit maximization, can be formulated as follows:

$$\max U_{\text{WP}} - C_{\text{WP}} \quad (9)$$

3.2. Model of PVPP

PVPP generates profits U_{PV} through transactions with ES and the power grid:

$$U_{\text{PV}} = U_{\text{PV2ES}} + U_{\text{PV2PG}} \quad (10)$$

$$U_{\text{PV2ES}} = \sum_{t=1}^{T} p_{\text{PV2ES}}^t P_{\text{PV2ES}}^t \quad (11)$$

$$U_{\text{PV2PG}} = \sum_{t=1}^{T} p_{\text{PV2PG}}^t P_{\text{PV2PG}}^t \quad (12)$$

where U_{PV2ES} and U_{PV2PG} are the profits made from the transaction with ES and the power grid, respectively; p_{PV2ES}^t is the trading price with ES at time t; p_{PV2PG}^t is the PV power feed-in tariff at time t; and P_{PV2ES}^t and P_{PV2PG}^t are the electricity sold to ES and the power grid, respectively [30].

The costs associated with PVPP C_{PV} include the maintenance fee, C_{PVM} and the network fee, C_{PVN}:

$$C_{\text{PV}} = C_{\text{PVM}} + C_{\text{PVN}} \quad (13)$$

$$C_{\text{PVM}} = \sum_{t=1}^{T} \omega_{\text{PV}} P_{\text{PV}}^t \quad (14)$$

$$C_{\text{PVN}} = \sum_{t=1}^{T} [\alpha_{\text{PV}} (P_{\text{PV2ES}}^t)^2 + \beta_{\text{PV}} P_{\text{PV2ES}}^t] \quad (15)$$

where ω_{PV} is the coefficient of the PVPP maintenance fee; P_{PV}^t is the electricity produced by PVPP at time t; and α_{PV} and β_{PV} are the network fee coefficients of PVPP.

The electricity produced by PVPP should satisfy the power balance constraint and should be less than the maximum electricity generation of PVPP, as follows:

$$P_{\text{PV}}^t = P_{\text{PV2ES}}^t + P_{\text{PV2PG}}^t \quad (16)$$

$$0 \le P_{\text{PV}}^t \le P_{\text{PV}}^{\max} \quad (17)$$

where $P_{\text{PV}}^{\max}$ is the maximum electricity generation of PVPP.

The objective function for PVPP, aimed at profit maximization, can be formulated as follows:

$$\max U_{PV} - C_{PV} \tag{18}$$

3.3. Model of ES

The constraints that ES must adhere to include the state of the charge constraints, the charge/discharge power constraints, and the ES multiplying factor constraints.

(1) State of charge constraints

$$E_{ES}^{t} = (1-\tau)E_{ES}^{t-1} + [\eta_{abs}P_{ES,abs}^{t} - \frac{1}{\eta_{relea}}P_{ES,relea}^{t}]\Delta t \tag{19}$$

$$P_{ES,abs}^{t} = \max\{0, -P_{ES}^{t}\} \tag{20}$$

$$P_{ES,relea}^{t} = \max\{0, P_{ES}^{t}\} \tag{21}$$

$$10\%E_{ES,max} \leq E_{ES}^{t} \leq 90\%E_{ES,max} \tag{22}$$

$$E_{ES}^{0} = 20\%E_{ES,max} \tag{23}$$

$$E_{ES}^{0} = E_{ES}^{T} \tag{24}$$

Here, τ denotes the self-discharge efficiency of ES; E_{ES}^{t} and E_{ES}^{t-1} denote the energy storage level of ES at time t and time $t-1$; η_{abs} and η_{relea} denote the charging and discharging efficiencies of ES; $P_{ES,abs}^{t}$ denotes the charging power of ES at time t; $P_{ES,relea}^{t}$ denotes the discharging power of ES at time t; P_{ES}^{t} denotes the charging and discharging power of ES at time t; and E_{ES}^{0} and E_{ES}^{T} denote the energy levels at the beginning and end of the ES operational period.

(2) Charge/discharge power constraint

$$-P_{ES,max} \leq P_{PG2ES}^{t} \leq P_{ES,max} \tag{25}$$

(3) ES multiplying factor constraints

The maximum capacity and maximum charge/discharge power of the ES are directly proportional. The specific are as follows:

$$E_{ES,max} = \beta P_{ES,max} \tag{26}$$

where β denotes the ES energy multiplication factor.

The costs associated with ES include the electricity cost and maintenance fee C_{ESM}:

$$C_{ES} = U_{WP2ES} + U_{PV2ES} + U_{PG2ES} + C_{ESM} \tag{27}$$

$$U_{PG2ES} = \sum_{t=1}^{T} p_{PG}^{t} P_{PG2ES}^{t} \tag{28}$$

$$C_{ESM} = \sum_{t=1}^{T} (\omega_{ES}P_{ES}^{t} + \omega_{LOAD}P_{LOAD}^{t}) \tag{29}$$

where U_{PG2ES} is the cost paid to the power grid; p_{PG}^{t} is the day-ahead market price at time t; P_{PG2ES}^{t} is the electricity purchased from the power grid at time t; ω_{ES} and ω_{LOAD} are the coefficients of the maintenance fee of the device and loads; and P^t_{LOAD} is the electricity demand of the load.

The operation of ES should satisfy the power balance constraint:

$$P^t_{\mathrm{WP2ES}} + P^t_{\mathrm{PV2ES}} + P^t_{\mathrm{PG2ES}} + P^t_{\mathrm{ES}} = P^t_{\mathrm{LOAD}} \tag{30}$$

The objective function for ES, aimed at profit maximization, can be formulated as follows:

$$\max - C_{\mathrm{ES}} \tag{31}$$

4. Nash Bargaining Theory

Nash bargaining theory is a fundamental concept in cooperative game theory that solves negotiation problems by determining the optimal and fair allocation of resources among multiple participants [35]. The theory is based on feasible sets, divergence points, and Nash bargaining solutions that aim to maximize the product of the participants' utilities relative to their respective divergence points. Despite its advantages, the Nash bargaining theory has its limitations. For example, in practice, the theory must take into account the effects and interactions of multiple factors [36]. In the context of energy dispatch, especially in multi-energy systems involving WP, PVPP, and ES, Nash bargaining theory offers significant advantages by optimizing cooperative operations, protecting privacy, and enhancing decision-making capabilities. Combining Nash bargaining with the power dispatch meta-universe platform can enhance the decision-making capabilities of power grids using advanced simulation and optimization techniques, enabling better management of renewable energy sources and ES, optimizing system performance, and ensuring privacy.

4.1. Nash Bargaining Model

In the multi-energy system constructed on the power dispatch meta-universe platform, WP, PVPP, and ES act as rational participants aiming to maximize their profits. They are incentivized to engage in cooperation if it leads to increased profits or reduced costs. As self-interested rational participants, WP, PVPP, and ES prioritize determining an optimal energy dispatch strategy to achieve a Pareto optimal solution, thereby maintaining their cooperative relationship. The Nash bargaining theory, an important component of cooperative game theory, proves effective in achieving such a solution. The standard Nash bargaining model is formulated as follows [30]:

$$\begin{cases} \max \prod\limits_{n=1}^{N} (U_n - U_n^0) \\ \text{s.t.} \quad U_n \geq U_n^0 \end{cases} . \tag{32}$$

where N is the number of participants; n is the index of participants; and U_n is the utility of the participant n.

Based on the above, the Nash bargaining model of the multi-energy system with WP, PVPP, and ES can be written as follows:

$$\begin{cases} \max \; (U_{\mathrm{WP}} - C_{\mathrm{WP}} - U^0_{\mathrm{WP}})(U_{\mathrm{PV}} - C_{\mathrm{PV}} - U^0_{\mathrm{PV}})(-C_{\mathrm{ES}} - U^0_{\mathrm{ES}}) \\ \text{s.t.} \;\; U_{\mathrm{WP}} - C_{\mathrm{WP}} \geq U^0_{\mathrm{WP}} \\ \qquad U_{\mathrm{PV}} - C_{\mathrm{PV}} \geq U^0_{\mathrm{PV}} \\ \qquad -C_{\mathrm{ES}} \geq U^0_{\mathrm{ES}} \end{cases} \tag{33}$$

where U^0_{WP}, U^0_{PV}, and U^0_{ES} are the rupture points of WP, PVPP, and ES.

4.2. Transformation of the Nash Bargaining Model

According to the basic inequality, the objective function of the Nash bargaining model (33) should satisfy the following:

$$\begin{aligned}&(U_{\mathrm{WP}}-C_{\mathrm{WP}}-U_{\mathrm{WP}}^{0})(U_{\mathrm{PV}}-C_{\mathrm{PV}}-U_{\mathrm{PV}}^{0})(-C_{\mathrm{ES}}-U_{\mathrm{ES}}^{0})\\&\leq\left(\frac{U_{\mathrm{WP}}-C_{\mathrm{WP}}+U_{\mathrm{PV}}-C_{\mathrm{PV}}-C_{\mathrm{ES}}-U_{\mathrm{WP}}^{0}-U_{\mathrm{PV}}^{0}-U_{\mathrm{ES}}^{0}}{3}\right)^{3}\\&=\left(\frac{U_{\mathrm{WP2PG}}-C_{\mathrm{WP}}+U_{\mathrm{PV2PG}}-C_{\mathrm{PV}}-U_{\mathrm{PG2ES}}-C_{\mathrm{ESM}}-U_{\mathrm{WP}}^{0}-U_{\mathrm{PV}}^{0}-U_{\mathrm{ES}}^{0}}{3}\right)^{3}\end{aligned}\tag{34}$$

The equality holds if and only if

$$(U_{\mathrm{WP}}-C_{\mathrm{WP}}-U_{\mathrm{WP}}^{0})=(U_{\mathrm{PV}}-C_{\mathrm{PV}}-U_{\mathrm{PV}}^{0})=(-C_{\mathrm{ES}}-U_{\mathrm{ES}}^{0})\tag{35}$$

Therefore, finding the solution of (33) is equivalent to find the solution of the following:

$$\max U_{\mathrm{WP2PG}}-C_{\mathrm{WP}}+U_{\mathrm{PV2PG}}-C_{\mathrm{PV}}-U_{\mathrm{PG2ES}}-C_{\mathrm{ESM}}-U_{\mathrm{WP}}^{0}-U_{\mathrm{PV}}^{0}-U_{\mathrm{ES}}^{0}\tag{36}$$

Since U_{WP}^{0}, U_{PV}^{0}, and U_{ES}^{0} are constants, the solution of (36) is equivalent to the solution of the problem of maximizing alliance profit:

$$\max U_{\mathrm{WP2PG}}-C_{\mathrm{WP}}+U_{\mathrm{PV2PG}}-C_{\mathrm{PV}}-U_{\mathrm{PG2ES}}-C_{\mathrm{ESM}}\tag{37}$$

By solving (37), the electricity that ES purchases from WP, PVPP, and the power grid can be obtained; i.e., the energy dispatch strategy can be obtained.

4.3. Distributed Solving Method Based on ADMM

Preserving the privacy of WP, PVPP, and the power grid in the multi-energy system is an important issue. The ADMM algorithm allows WP, PVPP, and ES to process data in a distributed manner without the need to centralize all the data to a single center for processing. On the one hand, the risk of data leakage that may be caused by centralized processing can be avoided. On the other hand, each participant can keep its private data and solve the problem locally with only intermediate data exchanged. Through that, the privacy of the participants can be preserved. Thus, ADMM is applied to solve (37).

New variables, $\dot{P}_{\mathrm{WP2ES}}^{t}$ and $\dot{P}_{\mathrm{PV2ES}}^{t}$, are introduced to denote the expected electricity purchased by ES from WP and PVPP, and P_{WP2ES}^{t} and P_{PV2ES}^{t} denote the expected electricity sold by WP and PVPP to ES [30]. When $\dot{P}_{\mathrm{WP2ES}}^{t}=P_{\mathrm{WP2ES}}^{t}$ and $\dot{P}_{\mathrm{PV2ES}}^{t}=P_{\mathrm{PV2ES}}^{t}$, WP and PVPP reach an agreement with ES on trading electricity.

The maximization problem is transferred to a minimization problem, and the augmented Lagrange function of (37) is formulated as follows:

$$\begin{aligned}&\min-(U_{\mathrm{WP2PG}}-C_{\mathrm{WP}}+U_{\mathrm{PV2PG}}-C_{\mathrm{PV}}-U_{\mathrm{PG2ES}}-C_{\mathrm{ESM}})+\sum_{t=1}^{T}\lambda_{\mathrm{WP2ES}}^{t}(\dot{P}_{\mathrm{WP2ES}}^{t}-P_{\mathrm{WP2ES}}^{t})\\&+\frac{\rho_{\mathrm{WP2ES}}}{2}\left\|\dot{P}_{\mathrm{WP2ES}}^{t}-P_{\mathrm{WP2ES}}^{t}\right\|_{2}^{2}+\sum_{t=1}^{T}\lambda_{\mathrm{PV2ES}}^{t}(\dot{P}_{\mathrm{PV2ES}}^{t}-P_{\mathrm{PV2ES}}^{t})+\frac{\rho_{\mathrm{PV2ES}}}{2}\left\|\dot{P}_{\mathrm{PV2ES}}^{t}-P_{\mathrm{PV2ES}}^{t}\right\|_{2}^{2}\end{aligned}\tag{38}$$

where $\lambda_{\mathrm{WP2ES}}^{t}$ and $\lambda_{\mathrm{PV2ES}}^{t}$ are Lagrange multipliers, and ρ_{WP2ES} and ρ_{PV2ES} are penalty factors.

Based on ADMM, the objective functions of WP, PVPP, and ES can be obtained by decomposing (38).

(1) WP

$$\begin{aligned}&\min-(U_{\mathrm{WP2PG}}-C_{\mathrm{WP}})+\sum_{t=1}^{T}\lambda_{\mathrm{WP2ES}}^{t}(\dot{P}_{\mathrm{WP2ES}}^{t}-P_{\mathrm{WP2ES}}^{t})+\frac{\rho_{\mathrm{WP2ES}}}{2}\left\|\dot{P}_{\mathrm{WP2ES}}^{t}-P_{\mathrm{WP2ES}}^{t}\right\|_{2}^{2}\\&\text{s.t. (1)–(8)}\end{aligned}\tag{39}$$

(2) PVPP

$$
\begin{aligned}
&\min -(U_{\mathrm{PV2PG}} - C_{\mathrm{PV}}) + \sum_{t=1}^{T} \lambda_{\mathrm{PV2ES}}^{t} (\dot{P}_{\mathrm{PV2ES}}^{t} - P_{\mathrm{PV2ES}}^{t}) + \frac{\rho_{\mathrm{PV2ES}}}{2} \left\| \dot{P}_{\mathrm{PV2ES}}^{t} - P_{\mathrm{PV2ES}}^{t} \right\|_2^2 \\
&\text{s.t. } (10)\text{–}(17)
\end{aligned}
\tag{40}
$$

(3) ES

$$
\begin{aligned}
&\min\,(U_{\mathrm{PG2ES}} + C_{\mathrm{ESM}}) + \sum_{t=1}^{T} \lambda_{\mathrm{WP2ES}}^{t} (\dot{P}_{\mathrm{WP2ES}}^{t} - P_{\mathrm{WP2ES}}^{t}) + \frac{\rho_{\mathrm{WP2ES}}}{2} \left\| \dot{P}_{\mathrm{WP2ES}}^{t} - P_{\mathrm{WP2ES}}^{t} \right\|_2^2 \\
&\quad + \sum_{t=1}^{T} \lambda_{\mathrm{PV2ES}}^{t} (\dot{P}_{\mathrm{PV2ES}}^{t} - P_{\mathrm{PV2ES}}^{t}) + \frac{\rho_{\mathrm{PV2ES}}}{2} \left\| \dot{P}_{\mathrm{PV2ES}}^{t} - P_{\mathrm{PV2ES}}^{t} \right\|_2^2 \\
&\text{s.t. } (19)\text{–}(30)
\end{aligned}
\tag{41}
$$

The steps to solve the Nash bargaining model are as follows:

(1) Initialization.

(2) ES: Based on $P_{\mathrm{WP2ES}}^{t}(k+1)$ and $P_{\mathrm{PV2ES}}^{t}(k+1)$, solve (41) to get $\dot{P}_{\mathrm{WP2ES}}^{t}(k)$ and $\dot{P}_{\mathrm{PV2ES}}^{t}(k)$.

(3) WP and PVPP: Based on $\dot{P}_{\mathrm{WP2ES}}^{t}(k+1)$ and $\dot{P}_{\mathrm{PV2ES}}^{t}(k+1)$, solve (39) and (40) to get $P_{\mathrm{WP2ES}}^{t}(k+1)$ and $P_{\mathrm{PV2ES}}^{t}(k+1)$.

(4) Update

$$
\lambda_{\mathrm{WP2ES}}^{t}(k+1) = \lambda_{\mathrm{WP2ES}}^{t}(k) + \rho_{\mathrm{WP2ES}} \left(\dot{P}_{\mathrm{WP2ES}}^{t}(k+1) - P_{\mathrm{WP2ES}}^{t}(k+1) \right) \tag{42}
$$

$$
\lambda_{\mathrm{PV2ES}}^{t}(k+1) = \lambda_{\mathrm{PV2ES}}^{t}(k) + \rho_{\mathrm{PV2ES}} \left(\dot{P}_{\mathrm{PV2ES}}^{t}(k+1) - P_{\mathrm{PV2ES}}^{t}(k+1) \right) \tag{43}
$$

(5) If

$$
\max\,\left(\sum_{t=1}^{T} \left\| \dot{P}_{\mathrm{WP2ES}}^{t}(k) - P_{\mathrm{WP2ES}}^{t}(k) \right\|_2^2, \sum_{t=1}^{T} \left\| \dot{P}_{\mathrm{WP2ES}}^{t}(k) - P_{\mathrm{WP2ES}}^{t}(k) \right\|_2^2 \right) < \delta \quad \text{or} \quad k > k_{\max} \tag{44}
$$

is satisfied, then (6). If not, $k = k + 1$, then go back to (2).

(6) Stop

Through the above steps, the energy dispatch strategy can be obtained with privacy preserved.

5. Case Studies

5.1. Case Setting

In this section, the proposed energy dispatch strategy is verified. The objective functions are solved by GUROBI in MATLAB 2023a.

Table 1 shows the relevant parameters of WP, PVPP, and ES.

Table 1. Parameters of WP and PVPP [30].

Parameters	Values	Parameters	Values
ω_{WP}	0.008	ω_{PV}	0.0085
α_{WP}	3×10^{-5}	α_{PV}	0.008
β_{WP}	0.01	β_{PV}	0.01
p_{WP2PG}	0.34 (CNY/kWh)	p_{PV2PG}	0.35 (CNY/kWh)
ω_{ES}	1.8×10^{-4}	ω_{LOAD}	0.022

5.2. Process of Distributed Solution

Figure 2 shows the iterative solution process of the cooperative operation problem of the multi-energy system with WP, PVPP, and ES based on the ADMM method. The iterative process stops at the 54th time, and it takes 72.21 s. It does not only solve the cooperative operation problem of the multi-energy system with WP, PVPP, and ES, but it also effectively preserves the privacy of the participants in a distributed manner.

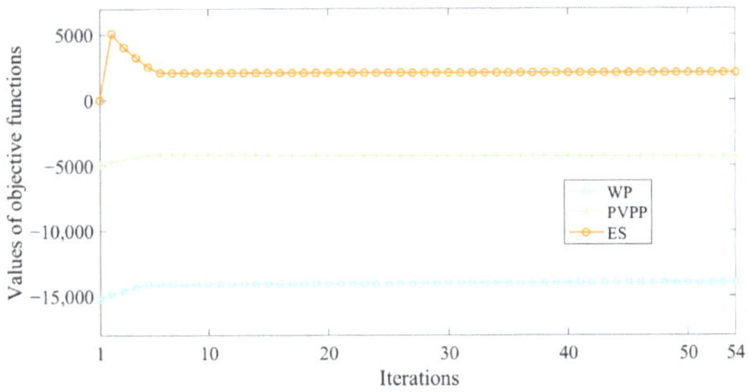

Figure 2. The iterative process for the cooperative operation problem.

5.3. Results Analysis

Figures 3 and 4 show the results of the ES's power purchased from WP, PVPP, and the grid and the results of the ES's operation in the power dispatch meta-universe platform, respectively. As can be seen from Figure 3, in order to reduce the cost of purchasing electricity, ES chooses to purchase 2303.67 kW from PVPP and 4421.30 kW from WP instead of the grid during the period of higher electricity price (8:00–22:00). While during the period of the lower electricity price (1:00–7:00) and (23:00–24:00), ES will choose to purchase 5317.25 kW of electricity from the grid instead of purchasing electricity from WP and PVPP. This allows the ES equipment to meet its own power demand, and the excess power is stored for later use during peak grid electricity prices, thus saving costs. The results show that ES purchases electricity from the grid during the grid electricity price valley periods, simultaneously meeting its own electricity demand and charging the ES device. The excess electricity is stored for later use during the grid electricity price peak periods, leading to cost savings. During periods of high grid electricity prices (8:00–22:00), ES chooses to purchase electricity from WP and PVPP. This can not only facilitate the consumption of renewable energy, but it also exhibits a notable peak-shaving effect. Additionally, ES further reduces the electricity cost paid the power grid by 4556.99 CNY. Additionally, the grid can yield an additional profit of 120.26 CNY by charging the network fees.

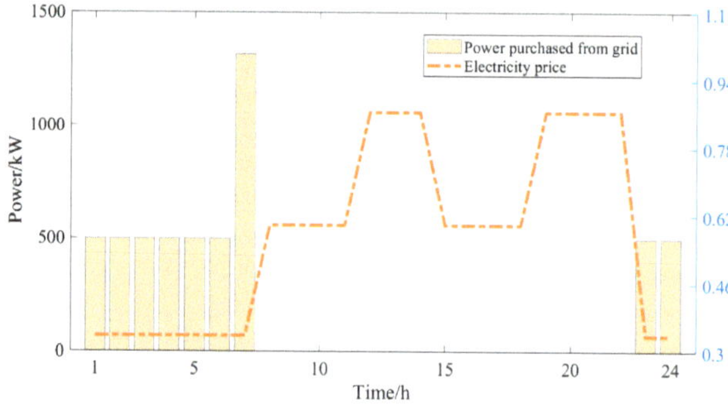

(**a**) Results of power purchased from the power grid by ES

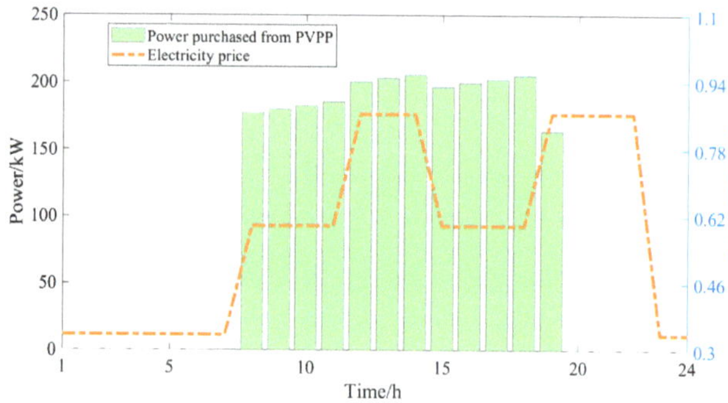

(**b**) Results of power purchased from PVPP by ES

Figure 3. *Cont.*

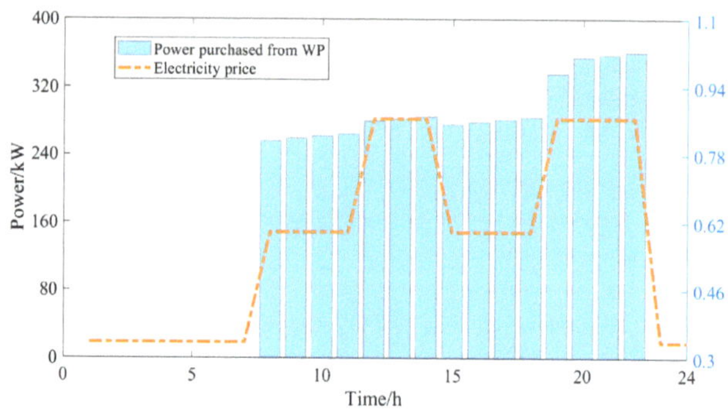

(**c**) Results of power purchased from WP by ES

Figure 3. Results of power purchased by ES.

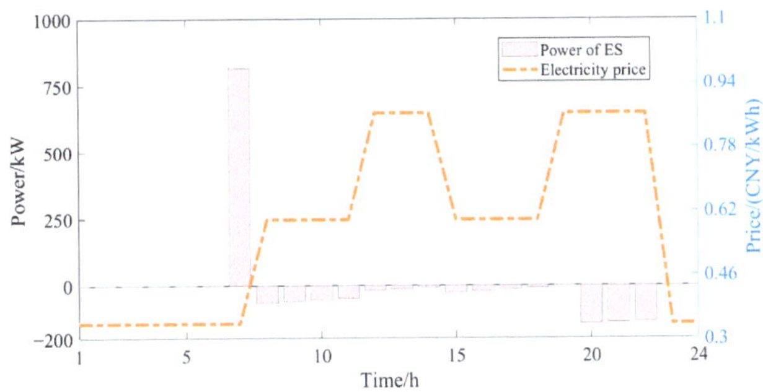

Figure 4. Operation results of ES.

Figure 5 shows the sale of electricity from WP and PVPP to the power grid. On the one hand, WP and PVPP can sell the affluent power to the grid to promote the consumption of new energy and increase the profits of WP and PVPP; on the other hand, the grid can reduce the pressure of power generation by purchasing power from WP and PVPP on the premise of meeting the demand for electricity. Of this, WP sells 42,426.10 kW to the grid and PVPP sells 12,848.93 kW to the grid.

In summary, the proposed energy dispatch strategy based on the power dispatch metauniverse achieves cooperative dispatching among WP, PVPP, and ES, demonstrates effective peak-shaving, facilitates the consumption of renewable energy sources, and contributes to the cost savings of ES.

Table 2 shows the overall profits in the case that WP, PVPP, and ES cooperate or not. The profit in the non-cooperation case is 14,111.28 CNY, and the profit in the cooperation case is 25,873.01 CNY. The comparison shows that the profit after the cooperation has increased by 11,761.73 CNY compared to the profit before the cooperation, which is an improvement of about 83.35%. This result verifies the effectiveness of the energy dispatch strategy proposed in this paper. In summary, the overall profit of the multi-energy system in the cooperation case is significantly improved by the energy dispatch strategy proposed in this paper.

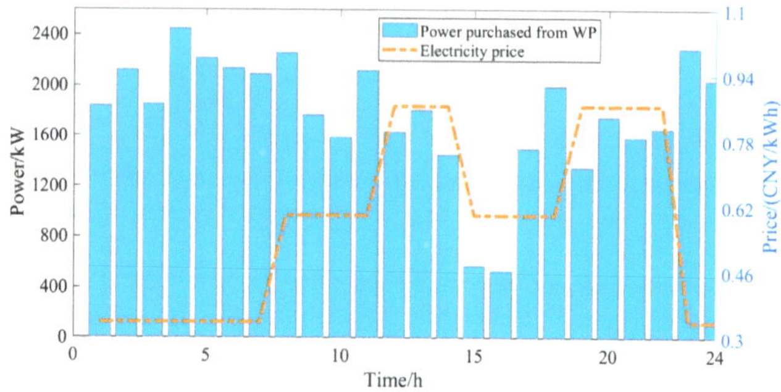

(**a**) Results of power purchased from WP by the power grid

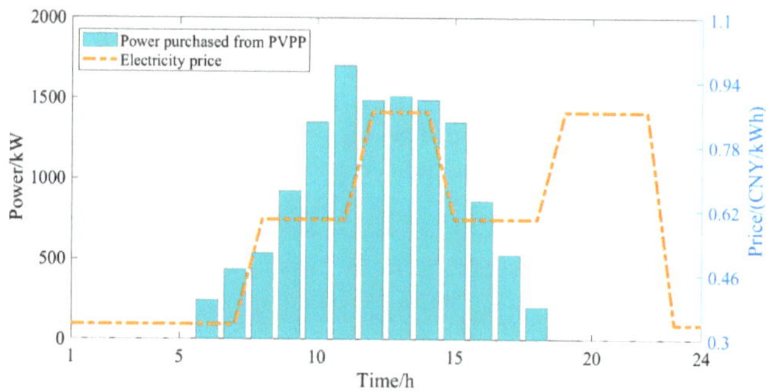

(**b**) Results of power purchased from PVPP by the power grid

Figure 5. Operation results of the power grid.

Table 2. Results of profits.

Cases	Alliance Profits (CNY)
Cooperation	25,873.01
Non-cooperation	14,111.28

6. Conclusions

This paper presents an intelligent distributed energy dispatch strategy for a multi-energy system integrating WP, PVPP, and ES. Modeling WP, PVPP, and ES within the power dispatch meta-universe platform facilitates a comprehensive depiction of their interrelationships and interactions. This framework enables distributed cooperative dispatching of power systems with a significant share of renewable energy, aiming to achieve optimal energy efficiency and economic benefits while ensuring privacy preservation. The main conclusions are as follows:

(1) The distributed optimization algorithm based on ADMM does not only realize the distributed and efficient solution of the cooperative operation problem of the multi-energy system, but it also effectively protects the privacy of WP, PVPP, and ES.

(2) The cooperative operation model established on Nash bargaining coordinates the dispatch of the multi-energy system through the power dispatch meta-universe platform, which promotes the consumption of renewable energy and has a certain peak-shaving effect for the grid.

(3) The validation of the energy dispatch strategy proposed in this paper through Nash bargaining based on the power dispatch meta-universe platform shows that the overall profit of the multi-energy system in the cooperation case is about 83.35% higher than that of the non-cooperation case.

(4) This method is helpful to realize the deduction of renewable energy dispatch strategies, and it further enhances the power grid intelligent decision-making ability of the power dispatching metaverse integrating multi-source spatiotemporal data.

Although the ADMM method effectively solves the dispatch problem of the multi-energy system with WP, PVPP, and ES in a distributed manner, the time to solve the problem is not satisfactory. Thus, the solving efficiency needs to be improved. In the future, we will improve the ADMM method to enhance the solving efficiency.

Author Contributions: J.L. and L.D.: conceptualization, methodology, validation, writing—original draft preparation, writing—review and editing, and supervision; J.C. and J.S.: methodology, validation, formal analysis and writing—original draft preparation; B.W. and Z.P.: conceptualization and supervision. All authors have read and agreed to the published version of the manuscript.

Funding: This work was funded by the Research on the Verification of Power Dispatch Metaverse Architecture and Key Basic Technologies (5108-202312052A-1-1-ZN).

Data Availability Statement: The data presented in this study are available on request from the corresponding author.

Conflicts of Interest: Author J.L. was employed by the company National Power Dispatching and Control Center, State Grid Corporation of China. Authors J.C. and J.S. were employed by State Grid Tianjin Electric Power Company. The remaining authors declare that the research was conducted in the absence of any commercial or financial relationships that could be construed as a potential conflict of interest.

References

1. Lu, R.; Chen, D.; He, H.; Zhang, S.; Zhang, J.; Zhang, Y.; Ma, L.; Wang, H. Balancing Supply and Demand in Hydropower, Wind, Solar, and Energy Storage Systems Using Heuristic Unit Combination Algorithm. *Water Resour. Power* **2023**, *41*, 222–226.
2. Zhang, H.; Zhai, X.; Zhang, J.; Bai, X.; Li, Z. Mechanism Analysis of the Effect of the Equivalent Proportional Coefficient of Inertia Control for a Doubly Fed Wind Generator on Frequency Stability in Extreme Environments. *Sustainability* **2024**, *16*, 4965. [CrossRef]
3. Jiang, M.; Wang, X.; Dong, C.; Wang, S.; Zhou, H.; Gao, J. Power Optimization Configuration of Water, Hydropower, Wind, and Solar Complementary System Based on Time-Sequential Simulation. *J. Hydroelectr. Eng.* **2023**, *1003–1243*, 1–11.
4. Mathiesen, B.V.; Lund, H.; Connolly, D.; Wenzel, H.; Østergaard, P.A.; Möller, B.; Nielsen, S.; Ridjan, I.; Karnøe, P.; Sperling, K.; et al. Smart Energy Systems for coherent 100% renewable energy and transport solutions. *Appl. Energy* **2015**, *145*, 139–154. [CrossRef]
5. Zakaria, A.; Ismail, F.B.; Lipu, M.H.S.; Hannan, M.A. Uncertainty models for stochastic optimization in renewable energy applications. *Renew. Energy* **2020**, *145*, 1543–1571. [CrossRef]
6. Yang, J.; Su, C. Robust optimization of microgrid based on renewable distributed power generation and load demand uncertainty. *Energy* **2021**, *223*, 120043. [CrossRef]
7. Pinson, P.; Madsen, H. Benefits and challenges of electrical demand response: A critical review. *Renew. Sustain. Energy Rev.* **2014**, *39*, 686–699.
8. Zhang, H.; Li, Z.; Xue, Y.; Chang, X.; Su, J.; Sun, H. A Stochastic Bi-level Optimal Allocation Approach of Intelligent Buildings Considering Energy Storage Sharing Services. *IEEE Trans. Consum. Electron.* **2024**. [CrossRef]
9. Koohi-Fayegh, S.; Rosen, M.A. A review of energy storage types, applications and recent developments. *J. Energy Storage* **2020**, *27*, 101047. [CrossRef]
10. Wang, W.; Yuan, B.; Sun, Q.; Wennersten, R. Application of energy storage in integrated energy systems—A solution to fluctuation and uncertainty of renewable energy. *J. Energy Storage* **2022**, *52*, 104812. [CrossRef]
11. ALAhmad, A.K.; Verayiah, R.; Shareef, H. Long-term optimal planning for renewable based distributed generators and battery energy storage systems toward enhancement of green energy penetration. *J. Energy Storage* **2024**, *90*, 111868. [CrossRef]
12. Guerra, O.J.; Zhang, J.; Eichman, J.; Denholm, P.; Kurtza, J. The value of seasonal energy storage technologies for the integration of wind and solar power. *Energy Environ. Sci.* **2020**, *13*, 1909. [CrossRef]
13. Guelpa, E.; Bischi, A.; Verda, V.; Chertkov, M.; Lund, H. Towards future infrastructures for sustainable multi-energy systems: A review. *Energy* **2019**, *184*, 2–21. [CrossRef]
14. Mancarella, P. MES (multi-energy systems): An overview of concepts and evaluation models. *Energy* **2014**, *65*, 1–17. [CrossRef]
15. Guo, Q.; Lu, Y. Key Technologies and Prospects of Modeling and Simulation of New Power System. *Autom. Electr. Power Syst.* **2022**, *46*, 18–32.
16. Lee, L.H.; Braud, T.; Zhou, P.; Wang, L.; Xu, D. All One Needs to Know about Metaverse: A Complete Survey on Technological Singularity, Virtual Ecosystem, and Research Agenda. *arXiv* **2021**, arXiv:2110.05352.

17. Zhao, S.; Gan, A.; Zhou, S.; Jian, X.; Zhang, Z. Comprehensive Energy Optimization Scheduling Based on Stackelberg Leader-Follower Game Multi-Objective Model. *J. Electr. Eng.* **2023**, *18*, 341–347.
18. Wei, F.; Jing, Z.; Wu, Z.; Wu, Q. A Stackelberg game approach for multiple energies trading in integrated energy systems. *Appl. Energy* **2017**, *200*, 315–329. [CrossRef]
19. Bao, T.; Zhang, X.; Yu, T.; Liu, X.; Wang, D. Stackelberg Game Model Reflecting Real-Time Supply-Demand Interaction and Its Reinforcement Learning Solution. *Proc. CSEE* **2018**, *38*, 2947–2955.
20. Liu, H.; Li, J.; Ge, S.; Zhang, P.; Chen, X. Coordinated Scheduling of Grid-connected Integrated Energy Microgrid Based on Multi-agent Game and Reinforcement Learning. *Autom. Electr. Power Syst.* **2019**, *43*, 40–48.
21. Song, X.; Chen, J.; Shi, Q. Optimal Scheduling of Integrated Energy Microgrids in a Multi-Agent Game Context. *High Volt. Eng.* **2023**, *49*, 3163–3178.
22. Cremers, S.; Robu, V.; Hofman, D.; Naber, T.; Zheng, K.; Norbu, S. Efficient methods for approximating the Shapley value for asset sharing in energy communities. In Proceedings of the Thirteenth ACM International Conference on Future Energy Systems, Virtual, 28 June–1 July 2022; pp. 320–324.
23. Chen, C.; Liu, C.; Ma, L.; Chen, T.; Wei, Y.; Qiu, W.; Li, Z. Cooperative-game-based joint planning and cost allocation for multiple park-level integrated energy systems with shared energy storage. *J. Energy Storage* **2023**, *73*, 108861. [CrossRef]
24. Wang, H.; Li, K.; Zhang, C.; Chen, J. Capacity and operation joint optimization for integrated energy system based on Nash bargaining game. *Energy* **2024**, *305*, 132032. [CrossRef]
25. Wang, Y.; Zheng, Y.; Yang, Q. Nash bargaining based collaborative energy management for regional integrated energy systems in uncertain electricity markets. *Energy* **2023**, *269*, 126725. [CrossRef]
26. Wang, Z.; Chen, L.; Li, X.; Mei, S. A Nash bargaining model for energy sharing between micro-energy grids and energy storage. *Energy* **2023**, *283*, 129065. [CrossRef]
27. Li, G.; Li, Q.; Song, W.; Wang, L. Incentivizing distributed energy trading among prosumers: A general Nash bargaining approach. *Int. J. Electr. Power Energy Syst.* **2021**, *131*, 107100. [CrossRef]
28. Chen, Y.; Pei, W.; Ma, T.; Xiao, H. Asymmetric Nash bargaining model for peer-to-peer energy transactions combined with shared energy storage. *Energy* **2023**, *278*, 127980. [CrossRef]
29. Ding, B.; Li, Z.; Li, Z.; Xue, Y.; Chang, X.; Su, J.; Sun, H. A CCP-based distributed cooperative operation strategy for multi-agent energy systems integrated with wind, solar, and buildings. *Appl. Energy* **2024**, *365*, 123275. [CrossRef]
30. Ma, T.; Pei, W.; Xiao, H.; Li, D.; Lv, X.Y.; Hou, K. Cooperative operation method for wind-solar-hydrogen multi-agent energy system based on Nash bargaining theory. *Proc. CSEE* **2021**, *41*, 25–39.
31. Jiang, D.; Guo, Y.; Liang, K.; Niu, J.; Shi, K.; Yang, C. Cooperative Game in a Multi-Electricity-Gas Interconnected System Considering New Energy Integration and Low Carbon Emissions. *Acta Energ. Sol. Sin.* **2023**, *44*, 58–67.
32. Zhang, X.; Wang, L.; Huang, L.; Wang, S.; Wang, C.; Guo, C. Integrated Energy Park Optimization Scheduling Considering Expanded Carbon Emission Flow and Carbon Trading Bargaining Model. *Autom. Electr. Power Syst.* **2023**, *47*, 34–46.
33. Jiang, A.; Yuan, H.; Li, D. Energy management for a community-level integrated energy system with photovoltaic prosumers based on bargaining theory. *Energy* **2021**, *225*, 120272. [CrossRef]
34. Mu, C.; Ding, T.; Qu, M.; Zhou, Q.; Li, F.; Shahidehpour, M. Decentralized optimization operation for the multiple integrated energy systems with energy cascade utilization. *Appl. Energy* **2020**, *280*, 115989. [CrossRef]
35. Wang, L.L.; Xian, R.C.; Jiao, P.H.; Liu, X.H.; Xing, Y.W.; Wang, W. Cooperative operation of industrial/commercial/residential integrated energy system with hydrogen energy based on Nash bargaining theory. *Energy* **2024**, *288*, 129868. [CrossRef]
36. Nazari, A.; Keypour, R.; Amjady, N. Joint investment of community energy storage systems in distribution networks using modified Nash bargaining theory. *Appl. Energy* **2021**, *301*, 117475. [CrossRef]

electronics

MDPI

Article

Game-Theory-Based Design and Analysis of a Peer-to-Peer Energy Exchange System between Multi-Solar-Hydrogen-Battery Storage Electric Vehicle Charging Stations

Lijia Duan [1], Yujie Yuan [2,*], Gareth Taylor [1] and Chun Sing Lai [1,*]

[1] Department of Electronic and Electrical Engineering, Brunel University London, London UB8 3PH, UK; lijia.duan@brunel.ac.uk (L.D.); gareth.taylor@brunel.ac.uk (G.T.)

[2] School of Air Traffic Management, Civil Aviation University of China, Tianjin 300300, China

* Correspondence: yjyuan@cauc.edu.cn (Y.Y.); chunsing.lai@brunel.ac.uk (C.S.L.)

Abstract: As subsidies for renewable energy are progressively reduced worldwide, electric vehicle charging stations (EVCSs) powered by renewable energy must adopt market-driven approaches to stay competitive. The unpredictable nature of renewable energy production poses major challenges for strategic planning. To tackle the uncertainties stemming from forecast inaccuracies of renewable energy, this study introduces a peer-to-peer (P2P) energy trading strategy based on game theory for solar-hydrogen-battery storage electric vehicle charging stations (SHS-EVCSs). Firstly, the incorporation of prediction errors in renewable energy forecasts within four SHS-EVCSs enhances the resilience and efficiency of energy management. Secondly, employing game theory's optimization principles, this work presents a day-ahead P2P interactive energy trading model specifically designed for mitigating the variability issues associated with renewable energy sources. Thirdly, the model is converted into a mixed integer linear programming (MILP) problem through dual theory, allowing for resolution via CPLEX optimization techniques. Case study results demonstrate that the method not only increases SHS-EVCS revenue by up to 24.6% through P2P transactions but also helps manage operational and maintenance expenses, contributing to the growth of the renewable energy sector.

Keywords: electric vehicle charging station; photovoltaic; hydrogen storage system; battery storage; peer-to-peer energy trading; game theory

check for updates

Citation: Duan, L.; Yuan, Y.; Taylor, G.; Lai, C.S. Game-Theory-Based Design and Analysis of a Peer-to-Peer Energy Exchange System between Multi-Solar-Hydrogen-Battery Storage Electric Vehicle Charging Stations. *Electronics* **2024**, *13*, 2392. https://doi.org/10.3390/electronics13122392

Academic Editors: Hanqing Yang, Jiajia Yang, Zening Li and Zhengmao Li

Received: 17 May 2024
Revised: 16 June 2024
Accepted: 17 June 2024
Published: 19 June 2024

1. Introduction

Electric vehicles (EVs) are at the forefront of the global shift towards sustainable transportation, offering a cleaner, more energy-efficient alternative to traditional engine vehicles [1–3]. EVs leverage advances in battery technology, electric motors, and power electronics to provide a driving experience that is not only environmentally friendly but increasingly competitive in terms of performance, range, and cost. As governments worldwide implement policies to reduce carbon emissions and consumers become more eco-conscious, the adoption of EVs is accelerating, supported by expanding infrastructure for charging and a growing recognition of their role in mitigating climate change [4,5]. Embracing the future of mobility, EVs represent a key component in the transition to a low-carbon economy, promising a greener, more sustainable future for transportation.

Amid the swiftly advancing energy transition, multi-energy electric vehicle charging stations (EVCSs) are emerging as crucial infrastructure components for promoting sustainable transportation. These EVCSs, which harness various renewable energy technologies, including solar, wind, hydrogen, and battery storage, provide versatile charging solutions that help stabilize grid loads and optimize energy utilization. Equipped with advanced management systems, multi-energy EVCSs can dynamically adjust charging power and methods in response to real-time variations in energy supply and demand. This capability not only boosts energy efficiency and cuts operational costs but also significantly supports the ecological transformation of the power grid [6]. As trailblazers in the new era of mobility, EVs play a key role in the shift towards a low-carbon economy, heralding a more sustainable and greener future for transportation.

The shift toward sustainable energy systems is gaining momentum, and in this evolving situation, EVCSs equipped with diverse energy options—such as solar energy, hydrogen storage systems, and battery storage—are becoming increasingly critical. The integration of these varied energy sources necessitates the development of innovative management strategies to ensure efficient operation, economic viability, and reliability of the power supply. Within this framework, game theory and mixed-integer linear programming (MILP) are being effectively utilized to design and analyze peer-to-peer (P2P) energy exchange systems. These analytical approaches help optimize the distribution and utilization of resources, ensuring that EVCSs can meet demand flexibly and sustainably.

Figure 1 shows a solar-hydrogen-storage-integrated electric vehicle charging station (SHS-EVCS), which utilizes the combined capabilities of photovoltaic panels, a hydrogen storage system, and battery storage to charge electric vehicles. This station features a solar array that captures sunlight and transforms it into electrical energy. This energy can either be used directly to charge EVs or be fed into the grid and stored in battery reserves for later use. For hydrogen energy storage, the system incorporates an electrolyzer that splits water molecules into hydrogen and oxygen. The hydrogen is then compressed and stored, ready to be used when needed. During peak demand periods, this hydrogen is converted back into electricity through a fuel cell to power EVs. Additionally, a battery storage unit captures any excess energy from the solar array or the fuel cell, ensuring a consistent energy supply to charge EVs even when the primary renewable sources are inactive.

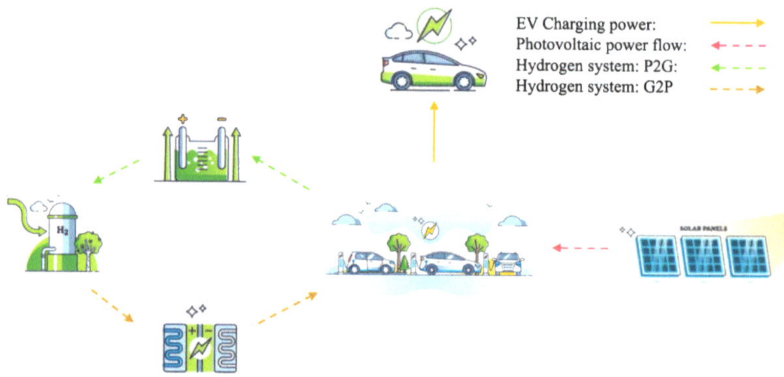

Figure 1. Single solar-hydrogen-storage-integrated electric vehicle charging station topology.

This paper mainly addresses the integration of renewable energy sources, energy management and optimization, and economic feasibility via game-theory-based P2P energy trading. The research focuses on the optimal sizing and integration of multiple renewable energy sources, including solar, hydrogen, and battery storage, within EV charging stations. This is crucial to maximize the utilization of renewable energy sources, minimize dependency on the grid, and ensure a sustainable and resilient energy supply. The combination of solar, hydrogen, and battery storage aims to leverage the complementary nature of these resources. Solar energy provides daytime power, hydrogen storage offers long-term energy storage, and batteries manage short-term fluctuations. This hybrid approach addresses the intermittent nature of renewable energy sources and improves the reliability and efficiency of the charging stations. This paper also develops advanced energy management strategies to optimize the operation of the SHS-EVCS. This includes the efficient allocation of energy resources, minimizing operational costs, and ensuring the reliability of the energy supply to EVs. By focusing on the economic aspects, the research aims to make SHS-EVCSs financially viable, which involves exploring energy trading opportunities among charging stations.

2. Literature Review

Over the past few decades, policy-driven and technology-driven changes have played a significant role, which is also reflected in the development of charging infrastructure and electric vehicles. The concept of electric vehicles, which dates back to the early 20th century, was first introduced alongside internal combustion engines. However, electric vehicles struggled to compete due to scarce charging options and limited range [7,8]. Electric vehicles did not have a high degree of popularity, and by the end of the 20th century, people's interest in electric mobility was stimulated by advances in electric battery technology and environmental issues [9]. Judging from the development in recent years, the progress made by electric locomotives is very significant in terms of electric vehicle charging infrastructure, especially in the integration process of online personnel. The role played by wind and solar charging stations is becoming more common, and sustainable alternatives can be provided by conventional electricity.

MILP, on the other hand, offers a method for optimizing the power distribution in this P2P network, enabling the charging stations to make decisions that collectively enhance the system's performance. This mathematical approach helps determine the most efficient allocation of resources, minimizing costs and maximizing the utilization of renewable energy.

P2P energy trading is a revolutionary approach in the energy sector that allows individuals or businesses to buy and sell electricity directly with each other, typically without the involvement of traditional energy suppliers. This model leverages technology like blockchain and smart grids to facilitate these transactions. P2P energy trading can lead to more efficient use of renewable energy, lower energy costs, and reduced dependence on large-scale utility companies. It empowers consumers to become 'prosumers'—energy producers and consumers—and can contribute to a more sustainable energy system. Zhou's study [10] explores P2P energy sharing in smart communities, focusing on its role in advancing renewable energy adoption. It reviews P2P systems, addresses challenges, and examines artificial intelligence- and blockchain-based strategies for efficient energy trading, highlighting the potential economic and operational benefits [10]. The study emphasizes the need for further research to enhance P2P system effectiveness in renewable energy markets, particularly focusing on the dynamics of P2P energy trading in community microgrids. The research develops a game-theory-based decentralized trading scheme that focuses on the impact of distributed energy resource (DER) ownership [11]. P2P trading benefits participants economically but can lead to losses in communities with high photovoltaic penetration. It underscores the need for strategic DER management to optimize economic outcomes in P2P energy trading [8]. In [12], the authors employ fuzzy optimization techniques to balance economic and environmental objectives in energy trading and propose a multi-period P2P trading model that simultaneously minimizes electricity costs and carbon emissions. Another paper uses a distributionally robust optimization approach, utilizing a fuzzy set based on Wasserstein distance for renewable energy prediction error, and proposes a day-ahead microgrid P2P transactive energy trading model, employing a linear and convex programming approach to address the nonlinear aspects of the model [13]. A novel model for optimizing P2P energy trading in multi-microgrid systems uses Nash bargaining theory and data-driven chance constraints to manage uncertainties in renewable energy and load forecasting [14]. In [15], the authors provide a comprehensive analysis of 50 global peer-to-peer distributed renewable energy trading projects.

Game theory, a mathematical framework for analyzing strategic interactions among rational decision-makers, can be applied effectively in renewable energy trading. This application is particularly relevant in markets like P2P energy trading, where multiple participants (consumers, prosumers, utility companies) interact [16–18]. Game theory provides a framework for understanding the strategic interactions between independent charging stations, each acting as rational players seeking to optimize their outcomes by modeling the decision-making processes of these entities, which can predict their behavior in response to varying energy supply, demand, and pricing conditions. In P2P energy trading, game theory can optimize energy distribution among participants. This involves strategies for energy distribution that are efficient, cost-effective, and fair, considering the varying production capacities and needs of different participants. In [16], the authors aimed to achieve higher economic income and maintain multi-agent income equilibrium. By employing the finite improvement property and a variable-step iterative convergence method, the study ensures efficient and accurate convergence of the model [16]. The model's simulation application demonstrates improved energy utilization rates and increased economic profits. The article [17] proposes a novel transactive energy market model using blockchain technology and game theory. The study introduces a proof-of-reserve consensus mechanism for prosumer–consumer transactions, enhancing energy trading efficiency and privacy, and uses game-theoretic market rules to establish a sustainable energy generation and consumption balance while ensuring economic agent privacy [17]. Another study [18] introduces a game-theory-based demand response program (DRP), integrating both incentive- and price-based DRP concepts, targeting residential, commercial, and industrial sectors; it evaluates three pricing strategies: fixed pricing, time-of-use pricing, and real-time pricing, along with their combinations. This approach leads to increased utility profits, reduced customer costs, and a more balanced load curve.

The fusion of game theory and MILP for designing and analyzing a P2P energy exchange system between multi-storage EV charging stations presents a novel solution that could revolutionize energy management in smart grids. This integrated approach addresses the economic aspects and contributes to the robustness and sustainability of energy systems in the era of decarbonization. This paper delves into such a system's complexities, exploring its potential to create a cooperative network that ensures energy security, fosters renewable integration, and paves the way for a greener future. It introduces a novel mixed-integer linear two-stage approach for analyzing the resiliency of power distribution systems. The study emphasizes the importance of considering both topological and electrical characteristics of distribution systems in resiliency analysis [19]. Study [20] introduces a MILP model optimized through a two-stage framework. This framework integrates system design with control problems of EV charging stations, using a design and analysis of a computer-experiments-based method [20]. Another study employs an iterative integer linear programming-based heuristic that efficiently tackles the multiple-choice knapsack problem with setup constraints [21]. This heuristic outperforms existing algorithms in solution quality and computation time, achieving optimal or near-optimal solutions for all tested instances. Linear programming and bi-objective optimization are particularly effective in long- and medium-term forecasts, reducing maximum errors significantly [22]. In Teng's paper [23], a distributed dual decomposition MILP-based energy management strategy for port-integrated energy systems is proposed to ensure reliable seaport operations.

Table 1 summarizes the model development and P2P using game theory. The paper [20] uses the CPLEX tool to solve the control problem of EVCSs. The study [24] uses several electricity sources to increase efficiency through game theory and P2P. Based on the previous studies, incorporating game theory into P2P systems involves using mathematical models to analyze and optimize interactions among participants with potentially conflicting interests. This approach is efficient in P2P networks, where decentralized entities negotiate, share resources, or trade without a central authority. Game theory provides a structured framework to model these interactions as strategic games, where each participant (player) seeks to maximize their utility through their actions [25–27].

Table 1. Literature review and benchmarking of relevant work.

Reference	Electricity Sources	Energy Exchange Applied	P2P or Game Theory Applied	Uses CPLEX Solver	Results
[20]	N/A	No	Yes	Yes	This approach utilizes a CPLEX solver tool to merge system design with control issues at EV charging stations, applying a method based on the design and analysis of computer experiments.
[28]	PV, grid	No	Yes	No	Using the suggested algorithm, prosumers enjoy increased earnings, while consumers experience reduced electricity expenses.
[29]	Grid, PV, wind	Yes	Yes	No	The suggested vehicle-to-vehicle market strategy enables EV owners to engage in peer-to-peer transactions, effectively lowering the cost of charging for individuals and evening out the demand on the power grid.
[30]	Grid	No	Yes	No	The proposed EV charging strategy optimally serves the individual interests of each EV, while also considering the demands of other EVs in the market for the next day.
[31]	Battery, PV	Yes	Yes	No	The method that this article proposed can significantly bolster resilience by up to 80% and extend battery lifespan by 32–37%.
[24]	Battery storage, PV, grid	Yes	Yes	No	The outcomes of the proposed strategy demonstrate an enhancement in energy utilization efficiency by considering the effects of the power system within peer-to-peer energy trading scenarios.
[32]	PV, wind, energy storage system, grid	No	Yes	No	This research demonstrates that the suggested control strategy provides a viable and practical method for managing an autonomous distributed system as miniature microgrids within the context of the electricity market, considering the competitive, non-cooperative relationships among microgrids.

Table 1. *Cont.*

Reference	Electricity Sources	Energy Exchange Applied	P2P or Game Theory Applied	Uses CPLEX Solver	Results
[33]	PV, storage system (heating, cooling, and battery), grid	Yes	Yes	No	The model offers a pragmatic and effective means for assisting in the determination of equitable trading prices, while also furnishing valuable perspectives on the optimal design of energy infrastructures.
[34]	PV, wind, energy storage system	No	Yes	No	The market simulation employs a non-model-based, game theory approach, wherein participants adjust their strategies based on the expected returns measured from the market.
[35]	PV, battery energy storage, hydrogen storage, grid	Yes	No	No	This holistic strategy for energy optimization allows the station to accommodate the variable energy needs for charging electric vehicles, thereby reducing costs, and enhancing sustainability.
This work	PV, battery energy storage, hydrogen storage, grid	Yes	Yes	Yes	The method not only boosts SHS-EVCS revenue through P2P transactions but also helps manage operational and maintenance expenses, contributing to the growth of the renewable energy sector.

The specific contributions of this work are as follows:

- Proposal of a P2P optimal dispatch strategy rooted in game theory for SHS-EVCS, aimed at achieving greater economic returns by ensuring income equilibrium across multiple SHS-EVCSs.
- Design and implementation of a CPLEX solver specifically to solve the linear-based simulation, predominantly employed for addressing locational queries. This solver is resolved and subsequently integrated with an SHS-EVCHs model to enhance its applicability.
- This study examines the synergistic collaborations and operational dynamics among diverse stakeholders to elucidate the shared economic benefits, with a particular emphasis on methodologies reliant on SHS-EVSC for the facilitation of energy sharing and the optimization of economic dispatch.

3. Problem Formulation

3.1. Objective Function

To reduce anticipated costs, the objective function F of SHS-EVCS includes the cost of hydrogen energy storage, gas turbine costs, solar energy costs, grid costs, battery storage costs, and P2P costs, minus the revenue from selling electricity (1).

$$F = min\sum_{n\in N}^{1} F_n = min\sum_{n}^{N} \sum_{t}^{T} \left(C_{n,t}^{E_H} + C_{n,t}^{FC} + C_{n,t}^{grid} + C_{n,t}^{Pv} + C_{n,t}^{p2p} + C_{t}^{Bes} - C_{n,t}^{sell}\right) \quad (1)$$

where

$C_{n,t}^{E_H}$ is hydrogen cost;

$C_{n,t}^{FC}$ is fuel cell generator cost;
$C_{n,t}^{grid}$ is grid cost;
$C_{n,t}^{Pv}$ is solar cost;
$C_{n,t}^{p2p}$ is p2p trading cost;
C_{t}^{Bes} is battery energy storage;
$C_{n,t}^{sell}$ is sale revenue.

3.2. Hydrogen System Model

3.2.1. Electrolyzer Model

The water electrolysis hydrogen production process uses water as raw material, and its composition mainly includes devices such as the electrolyzer, hydrogen (oxygen) gas–liquid–solid separators, hydrogen (oxygen) gas coolers, and hydrogen (oxygen) gas purifiers [36]. Using water decomposition to produce hydrogen is an efficient, environmentally friendly, and convenient new method for hydrogen production. The mathematical model of the electrolyzer for hydrogen production adopted in this paper is shown in Equation (2):

$$E_{AE,t} = \frac{v}{a} P_{AE.t} \tag{2}$$

$P_{AE.t}$ and $E_{AE,t}$ represent the electrical power consumed and the quantity of hydrogen produced by the electrolyzer, respectively. v denotes the conversion efficiency, which is 75%, and a represents the conversion coefficient for electric energy to the equivalent energy in hydrogen, valued at 39.65 kWh/kg [37].

3.2.2. Fuel Cell Generator Model

Hydrogen fuel cells, as an energy conversion device within the system, are an important part of hydrogen energy applications. Hydrogen fuel cells differ from dry cells and batteries, which are types of energy storage devices that store energy and release it where needed [38]. Strictly speaking, hydrogen fuel cells are not energy storage devices but rather devices that can generate electricity through the chemical properties of hydrogen and oxygen. Their basic principle is the reversible reaction of water electrolysis. This paper focuses on the process of the battery generating electrical energy by consuming hydrogen. Therefore, the mathematical expression for the hydrogen fuel cell adopted in this article is shown as Equation (3):

$$P_{FC} = \eta_{FC} a E_{FC} \tag{3}$$

P_{FC} and E_{FC} represent the output power and the quantity of hydrogen consumed by the fuel cell, respectively. η_{FC} is the energy conversion efficiency, which is 70% [37].

3.2.3. Hydrogen Storage Tank Model

Hydrogen storage tanks [39], as devices for storing hydrogen, come in different types based on various hydrogen storage technologies. The technology currently used most is high-pressure gaseous hydrogen storage. The storage tanks used in this technology are mainly made of conventional stainless acid-resistant steel plates and aluminum alloys. The process technology for their design and production is relatively mature, with low cost, a rapid rate of gas charging and discharging, and the capability to store hydrogen at normal temperatures. The mathematical model for the high-pressure hydrogen storage tanks used in this paper can be represented by Equation (4):

$$E_{hs}(t) = (1-\delta_{hs})E_{hs}(t-1) + \left[E_{hs}^{in}(t)\eta_{hs}^{in} - \frac{E_{hs}^{out}(t)}{\eta_{hs}^{out}}\right]\Delta t \tag{4}$$

$E_{hs}(t-1)$ and $E_{hs}(t)$ represent the total quantity of hydrogen stored in the hydrogen tank at time $t-1$ and t, respectively; δ denotes the energy storage decay rate of the hydrogen tank, which is 0.05; $P_{hs}^{in}(t)$ represents the input quantity of hydrogen into the hydrogen tank; $P_{hs}^{out}(t)$ represents the output quantity of hydrogen from the hydrogen tank; η_{hs}^{in} is the input efficiency of the hydrogen tank, which is 0.98; and η_{hs}^{out} is the output efficiency of the hydrogen tank, also 0.98 [35].

3.3. Photovoltaic Model

Based on the photovoltaic effect, the PV cells in a PV power generation system can utilize sunlight to produce electrical energy. The power generation efficiency model is given as Equation (5):

$$P_{pv} = P_{STC}G_{AC}\frac{[1+k(T_c-T_r)]}{G_{STC}} \tag{5}$$

k is the power temperature coefficient, valued at -0.45; different photovoltaic components will have different values, generally taken as $-0.3\%/^{\circ}\mathrm{C}$; and P_{STC} is the rated output power of the PV components under standard test conditions. G_{AC} is the real-time solar irradiance intensity where the photovoltaic cell operates; T_c is the real-time temperature of the photovoltaic component surface; G_{STC} is the illumination under standard test conditions, valued at 1000 $\mathrm{W/m^2}$; and T_r is 25 °C.

$$T_C = T_a + \frac{G_C}{800}(T_N - 20) \tag{6}$$

T_a is the ambient temperature, and T_N is the rated temperature at which the photovoltaic cell operates.

3.4. Battery Energy Storage Model

Batteries can effectively smooth out fluctuations caused by sudden changes in load demand within a microgrid. When the output power of photovoltaics exceeds the load demand power, the battery is in a charging state; conversely, when the output power is less than the load demand power, the battery is in a discharging state. Its characteristics are shown in (7) and (8).

Discharging model:

$$E_{BSS}(c) = E_{BSS}(c-1) + [E_{gen}(c) - \left(\frac{E_{load}(c)}{e_{inv}}\right)]\eta_b^{dch} \tag{7}$$

Charging model:

$$E_{BSS}(c) = E_{BSS}(c-1) - [E_{gen}(c) - \left(\frac{E_{load}(c)}{e_{inv}}\right)]/\eta_b^{ch} \tag{8}$$

where $E_{BSS}(c)$ is the energy of the battery at the c-th time, $E_{gen}(c)$ is the energy generated by the generator at the c-th time, $E_{load}(c)$ is the energy of the load at the c-th time, e_{inv} is the efficiency of the inverter, valued at 95%, and η_b^{ch} and η_b^{dch} represent the charging and discharging efficiency of the battery, both valued at 80% [40].

3.5. P2P Trading Model

3.5.1. P2P Internal Trading Revenue [13]

$$C_{n,t}^{sell} = \rho_{n,t}^{inter} P_{n,t}^{load} \tag{9}$$

$\rho_{n,t}^{inter}$ is the price of number n SHS-EVCSs selling electricity to other SHS-EVCSs using internal load.

3.5.2. P2G Trading with Grid

$$C_{n,t}^{grid} = \rho_{n,t}^{B} P_{n,t}^{B} - \rho_{n,t}^{g} P_{n,t}^{g} \tag{10}$$

$\rho^{B}_{n,t}$and $P^{B}_{n,t}$ are the price and electricity purchased by SHS-EVCS n from shared battery energy storage at time t. $\rho^{g}_{n,t}$ and $P^{g}_{n,t}$ are the price and power sold to the grid. Usually, $\rho^{g}_{n,t}$ is much lower than $\rho^{B}_{n,t}$, which can increase the profits or reduce costs through P2P trading.

3.5.3. P2P Transaction Energy Trading Cost

$$C^{p2p}_{n,t} = \sum\nolimits^{1}_{i\in N\backslash n} \rho^{i}_{n,t} P^{i}_{n,t} \tag{11}$$

where N is the set of charging stations participating in P2P trading. $\rho^{i}_{n,t}$ is the transaction price between SHS-EVCS n and SHS-EVCS N at time t.

3.5.4. P2P Transactive Energy Trading Constraints

For any time, the electricity sold and purchased should be balanced.

$$P^{i}_{n,t} = -P^{n}_{i,t}, \forall t \in T, \forall n \in N, i \in N\backslash n \tag{12}$$

3.6. *Game Theory Model*

Cooperative game theory [41] is a branch of game theory that focuses on understanding the behavior and strategies of groups of players who can form coalitions and collaborate to achieve better outcomes than they might individually. Unlike non-cooperative game theory, where the emphasis is on individual players' strategies and equilibria, cooperative game theory examines how groups of players can work together and how the collective benefits can be distributed among them. Key elements include the characteristic function, which defines the potential payoff for any coalition, and solution concepts like the Core, the Shapley Value, and the Nash Bargaining Solution, which help determine fair ways to distribute the collective payoff among the members.

Gamer: N(SHS-EVCS1(1), SHS-EVCS2, SHS-EVCS3... SHS-EVCSN)

Strategy: electricity prices (ρ); power transaction value set (P)

Matrix:

$\rho = \left[\rho^{cs}_{1,1} \cdots \rho^{cs}_{N,t} \cdots \rho^{cs}_{n,t} \cdots \rho^{cs}_{N,t} \cdots \rho^{cs}_{N,T}\right]$ All gamers' pricing at different times.

$P = \left[P^{cs}_{1,1} \cdots P^{cs}_{N,t} \cdots P^{cs}_{n,t} \cdots P^{cs}_{N,t} \cdots P^{cs}_{N,T}\right]$ All gamers' transaction value.

Benefits: $U = \left[U^{cs}_{1,1} \cdots U^{cs}_{N,1} \cdots U^{cs}_{n,t} \cdots U^{cs}_{N,t} \cdots U^{cs}_{N,T}\right]$ Benefit set matrix consisting of the benefits of purchasing and selling electricity at different times for all players in the game.

Buyer: if there is energy exchange between charging stations, and the upper limit is $E^{g}_{n,t}$ (electricity buying from grid), then $\left|E^{cs}_{n,t}\right| \le E^{g}_{n,t}$; the SHS-EVCS is called a buyer at time t.

Seller: if there is energy exchange between charging stations, and the upper limit is $E^{g*}_{n,t}$ (electricity selling to grid), then $E^{cs}_{n,t} \le E^{g*}_{n,t}$; the SHS-EVCS is called a seller at time t.

When $SHS-EVCS_n \in Sellers$, the benefits at time t are

$$U^{cs}_{n,t} = \rho^{cs}_{n,t} P^{cs}_{n,t} \Delta t + \rho^{g}_{n,t}\left(E^{g*}_{n,t} - E^{cs}_{n,t}\right)\Delta t = \sum\nolimits_{i\in buyers} \rho^{cs}_{n,t} E^{cs}_{ni,t} \Delta t + \rho^{g}_{n,t}\left(E^{g*}_{n,t} - \sum\nolimits_{i\in buyers} E^{cs}_{ni,t}\right)\Delta t \tag{13}$$

When $SHS-EVCS_n \in Buyers$, the benefits at time t is:

$$U^{cs}_{i,t} = \sum\nolimits_{n\in sellers} \rho^{cs}_{n,t} E^{cs}_{in,t} \Delta t + \rho^{g*}_{n,t}\left(E^{g}_{n,t} - \sum\nolimits_{n\in sellers}\left|E^{cs}_{in,t}\right|\right)\Delta t \tag{14}$$

where $P^{cs}_{in,t}$ means *SHS-EVCS i* follows the *SHS-EVCS n* pricing rule $\rho^{cs}_{n,t}$ to buy the power at time *t*, $E^{cs}_{in,t} < 0$

3.7. Model Constraints

3.7.1. Hydrogen Constraints

- Electrolyzer Constraint:

$$u_{hs}(t)E^{o}_{hs,min} \le E^{o}_{ae}(t) \le u_{hs}(t)E^{o}_{hs,max} \tag{15}$$

where $E^{o}_{hs,min}$ is the lower output limit of the electrolyzer, $E^{o}_{hs,max}$ is the upper output limit of the electrolyzer, and $u_{hs}(t)$ is the state variable of the electrolyzer.

- Hydrogen tank constraints:

$$\begin{cases} u^{in}_{hs}(t)E^{in}_{hs,min} \le E^{in}_{hs}(t) \le u^{in}_{hs}(t)E^{in}_{hs,max} \\ u^{out}_{hs}(t)E^{out}_{hs,min} \le E^{out}_{hs}(t) \le u^{out}_{hs}(t)E^{out}_{hs,max} \\ E_{hs,min} \le E_{hs}(t) \le E_{hs,max} \\ u^{in}_{hs}(t) + u^{out}_{hs}(t) \le 1 \end{cases} \tag{16}$$

where $E^{in}_{hs,min}$ is the lower intput limit of the electrolyzer, $E^{in}_{hs,max}$ is the upper input limit of the electrolyzer, u^{in}_{hs} is the hydrogen tank storage state variable, and u^{out}_{hs} is the hydrogen tank release state variable.

- Fuel cell generator constraint:

$$0 \le P^{out}_{FC}(t) \le P^{out}_{FC,max} \tag{17}$$

where $P^{out}_{FC,max}$ is the maximum power output for the fuel cell generator.

3.7.2. Photovoltaic Constraint

$$P_{pv,\ min} < P_{PV}(t) < P_{pv,\ max} \tag{18}$$

where $P_{pv,min}$ and $P_{pv,max}$ are the minimum and maximum PV power output.

3.7.3. Battery Storage Constraint

$$E^{min}_{BSS} \le E_{BSS,j}(k) \le E^{max}_{BSS} \tag{19}$$

$$\begin{cases} E^{min}_{BSS} = (1 - DOD)E^{max}_{BSS} \\ E^{max}_{BSS} = N_{BSS}E_{rate\_BSS} \end{cases} \tag{20}$$

where E_{rate_BSS} is the self-discharge rate of the battery, and DOD is the depth of discharge, valued at 90%.

4. Case Study

Figure 2 provides a detailed schematic representation of four SHS-EVCSs engaged in energy exchange. The illustration uses a yellow line to denote the process of EV charging. The light blue bidirectional arrows symbolize the dynamic transactions of buying and selling electricity between the EVCSs, shared storage facilities, and the grid. Green dashed lines within the hydrogen storage system delineate the conversion of electrical power into gas via the electrolyzer, while the dark yellow dashed lines indicate the reconversion of gas back into electricity through the fuel cell generator. Additionally, pink dashed lines associated with the solar arrays illustrate photovoltaic power generation. A notable feature of this system is represented by the blue double-arrow line, which indicates the comprehensive energy exchanges occurring among the four SHS-EVCSs.

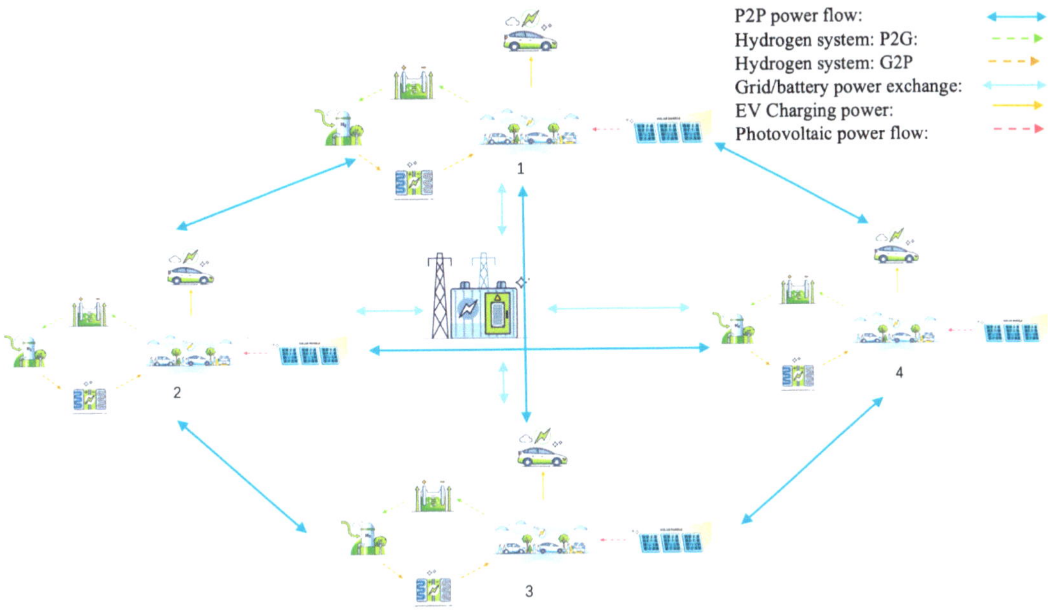

Figure 2. Topology of multi-SHS-EVCSs.

In the methodology simulation, Tables 2 and 3 detail the technical and economic parameters for four SHS-EVCSs, providing a comprehensive overview of each station's operational capabilities and financial aspects. These parameters include, but are not limited to, energy production capacities, operational efficiencies, maintenance and operational costs, and potential revenue streams. Table 4 presents a set of design variables crucial for the simulation process. These variables are instrumental in modeling the performance and economic viability of the stations under different scenarios, including energy demand fluctuations, market price variations, and changes in operational conditions.

Table 2. Technical parameters of SHS-EVCSs in four Dali boroughs [27,34,39].

Parameters	**Longzu 1 (EVCS1)**	**Longzu 2 (EVCS2)**	**Qiliqiao (EVCS3)**	**Fuyuan (EVCS4)**
Charger capacity (kW)	360	360	360	360
Number of chargers per station	10	10	12	8
PV installed capacity(kW)	1500	1500	2000	1000
Shared battery capacity (kWh)	10,000	10,000	10,000	10,000
Hydrogen tank capacity (m^3)	2000	2000	3000	1500
Fuel cell generator capacity (kW)	800	800	1000	600
Battery initial state of charge (%)	40	40	40	40
Minimum battery state of charge (%)	25	25	25	25
Maximum battery state of charge (%)	100	100	100	100
Battery charge and discharge efficiency (%)	80	80	80	80
Initial capacity of gas tank (%)	30	30	30	30
Tank storage efficiency (%)	98	98	98	98
Energy-to-gas efficiency (%)	70	70	70	70
Electricity-to-gas coefficient (kWh/m^3)	0.2	0.2	0.2	0.2
Hydrogen conversion efficiency (%)	75	75	75	75
Gas-to-electricity coefficient (m^3/kWh)	0.295	0.295	0.295	0.295

Table 3. Economic parameters of SHS-EVCSs in four Dali boroughs [42–44].

Parameters	Longzu 1	Longzu 2	Qiliqiao	Fuyuan
PV capital cost (GBP/kW)	286	286	286	286
Battery capital cost (GBP/kWh)	39.6	39.6	39.6	39.6
Hydrogen tank cost (GBP/m^3)	7.5	7.5	7.5	7.5

Table 4. Design variables for SHS-EVCSs [35,36,40].

Input	Technical Specification
PV output power [kW]	$P_{STC} G_{AC} \frac{[1+k(T_c-T_r)]}{G_{STC}}$
Hydrogen output power [kW]	$E^{t-1}_{H_2,i} - \left(P^t_{H-FC,i} + P^t_{SH,i} + P^t_{H_2,i}\right)\Delta t$
Battery output power [kW]	$P_{Bat,e,t1}(1-\sigma_{Bat,e}) + \left(P^{cha}_{Bat,e,t} * \eta^{cha}_{Bat,e} + \frac{P^{dis}_{Bat,e,t}}{\eta^{dis}_{Bat,e}}\right)$

Dali, located in Yunnan Province, China, was chosen as the case study primarily due to its geographic features and the prosperity of its tourism sector. Surrounded by mountains and lakes, Figure 3 shows that Dali's transportation is heavily dependent on two main north–south highways, and the constant tourism demand throughout the year further exacerbates transportation needs. Therefore, constructing charging stations has become a key measure to optimize the transportation infrastructure and support sustainable development. This not only helps alleviate traffic pressure but also encourages the use of eco-friendly transportation methods, such as electric vehicles, thereby enhancing Dali's tourism experience and environmental protection levels. There are numerous charging points in the parking area, but most require both parking and charging fees, which can be inconvenient for users. SHS-EVCSs offer a more cost-effective solution for electric vehicle owners. Additionally, SHS-EVCSs prioritize providing a safer and more secure charging environment, ensuring peace of mind for all EV drivers.

Figure 3. Topology of 2 main roads in Dali.

In the cooperative game model within the SHS-EVCS P2P trading framework, each EVCS functions as both a producer and a consumer, indicating that multiple entities reach a consensus. Therefore, the game-theory-based P2P process, as depicted in Figures 5–15, follows these steps and Figure 4 shows the game-theory-based P2P flow chart:

(a) Initiate trading.
(b) All SHS-EVCSs request energy exchanges.
(c) Aim to maximize the coalition's profits. Analyze each EVCS's energy trading needs to determine if they meet the model's constraints during the specified time (ensuring optimal profit for the coalition at time t), and then calculate the energy trading plan.
(d) Confirm the energy trading flow and calculate profits.
(e) If the profit maximizes the coalition's benefits, proceed to step (f); if not, return to step (c) and recalculate the energy exchange quantities.
(f) Allocate profits based on the constraints.
(g) Conclude the energy trading.

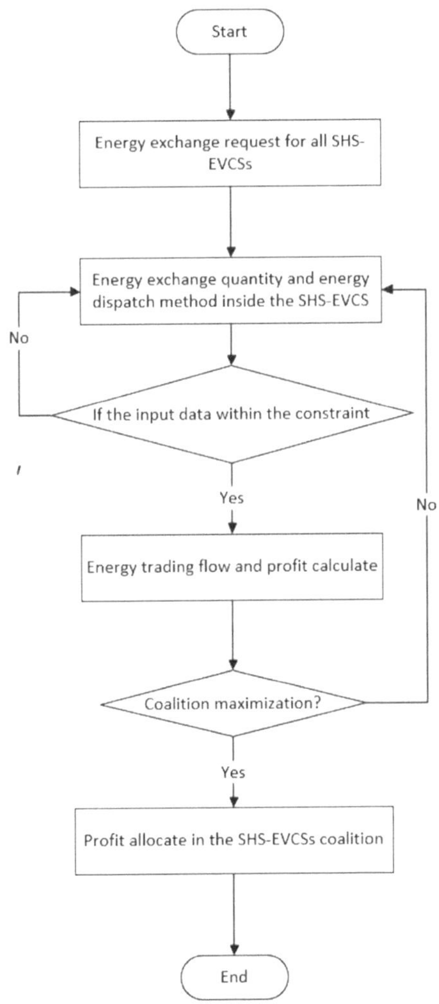

Figure 4. Game-theory-based P2P flow chart.

5. Results and Discussion

Figure 5 provides a detailed depiction of the variation in grid electricity prices and internal electricity trading prices over time. Grid electricity price remains essentially stable at 0.018 GBP/kWh until after 8 a.m., where it rises slightly to 0.05 GBP/kWh. This reflects the base price of the grid or time-specific tariff adjustments. The internal electricity purchase price briefly rises to 0.048 GBP/kWh around 7 a.m., then rises again around 0.075 GBP/kWh at 11, maintaining this level until 1 p.m. After 6 p.m., the purchase price rises once again to 0.075 GBP/kWh and continues until it starts to decline after 8 p.m. The internal electricity sale price rises to about 0.47 GBP/kWh around 7 a.m., then rises around 11 a.m., and drops again at 2 p.m., inversely mirroring the purchase price movements. After 6 p.m., the sale price sharply rises to 0.75 GBP/kWh, remaining at this level until it starts to decline after 8 p.m. These price fluctuations reflect changes in supply and demand within the internal market or adjustments in electricity trading strategies. Price increases represent an increase in demand or a decrease in supply, while decreases indicate a reduction in demand or an increase in supply. The disparity between the internal purchase and sale prices represents the margin for trading profits or opportunities for cost recovery.

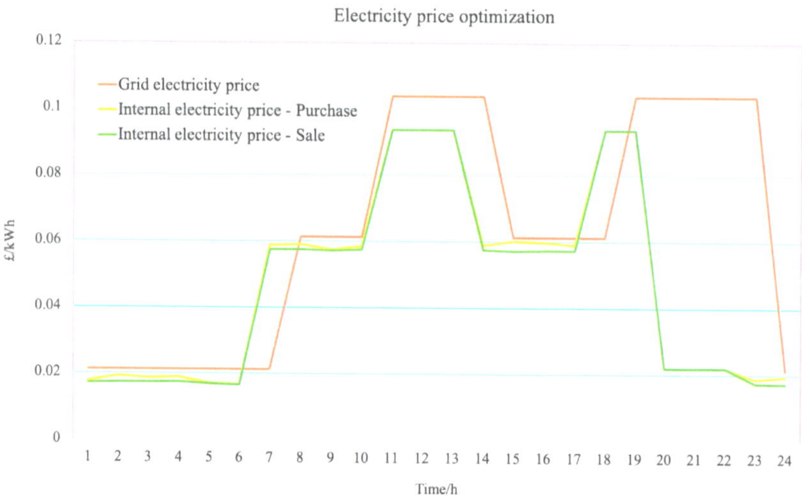

Figure 5. Electricity price optimization.

Figure 6 shows the number of electric vehicles for each charging station after optimization on different days. During the weekend, SHS-EVCS 1 shows an initial decrease after 2 a.m., followed by a slight drop after 6 a.m. A significant decrease is observed after 1 p.m., when the number of charges drops sharply. SHS-EVCS 2 exhibits that at 1 p.m., the first 25 cars charge in this station. At 8 p.m., 11 p.m., and 12 a.m., there are only 25 cars at each time. For SHS-EVCS 3, there are only 26 drivers who charge their cars at 6 p.m. SHS-EVCS 4, which has the most equal charging number for 24 h, only increased the charging number at 11 p.m. During weekdays, all EVCSs particularly show significant activity peaks around 9 a.m. and 10, correlating with typical commuting times. EVCS2 is the most utilized, suggesting higher demand or more frequent use compared to the others. All stations exhibit minimal activity during the early morning, which is before 6 a.m. This figure displays the traffic flow optimization results for four SHS-EVCSs. When designing this simulation model, multiple factors were considered, including geographical constraints, traffic control measures, and energy exchange. Through the analysis of these factors, the simulation model provided an optimal traffic flow management strategy. The increase and decrease indicated in the figure refer to the traffic flow adjustments at each charging station under specific conditions based on the optimal solutions derived from the simulation outputs. By adjusting traffic flows, we can more effectively distribute and utilize charging resources, thus addressing potential congestion issues during the electric vehicle charging process. Overall, this figure provides a comprehensive plan on how to optimally manage and schedule vehicle access to charging stations, ensuring charging efficiency while also enhancing the user's charging experience. Overall, this information is valuable for future research to optimize operations, perhaps by reducing power supply or staffing during low-demand hours.

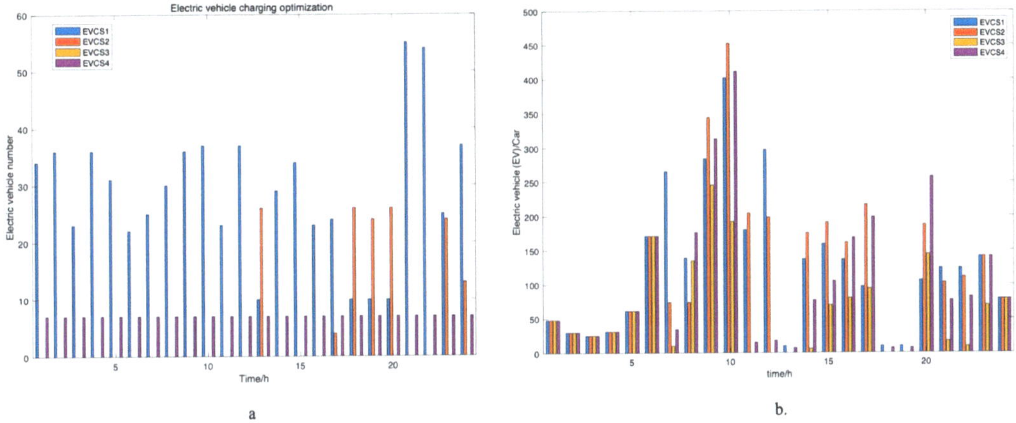

Figure 6. Optimization of the number of electric vehicle charging stations: (**a**) weekend; (**b**) weekday.

Figures 7–10 delineate the energy utilization and electricity load profiles of four distinct SHS-EVCSs. During the daytime, photovoltaic generation emerges as the predominant source of electricity, harnessing solar irradiance. Figures 9 and 10 show similar trends in energy consumption and electrical load, attributable to their analogous locational contexts, predominantly residential in nature. After sunset, the reliance transitions to hydrogen fuel cells for electricity generation, supplemented by acquisitions of electricity from a communal battery storage system. This shift underscores the versatility of the integrated energy systems within the SHS-EVCS framework.

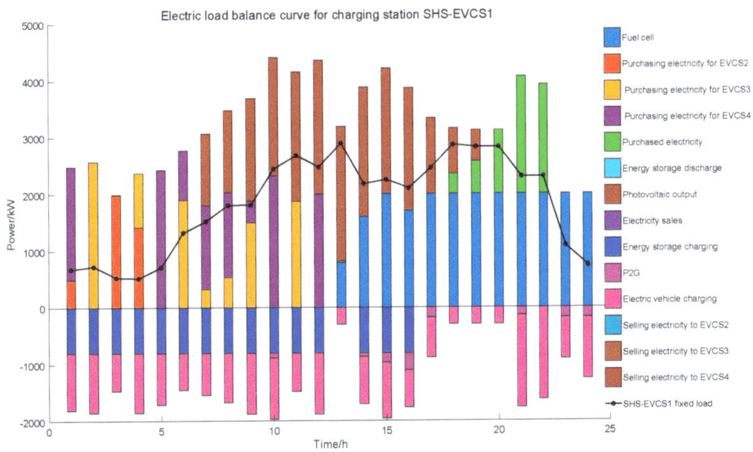

Figure 7. SHS-EVCS1 renewable energy usage and electricity load curve.

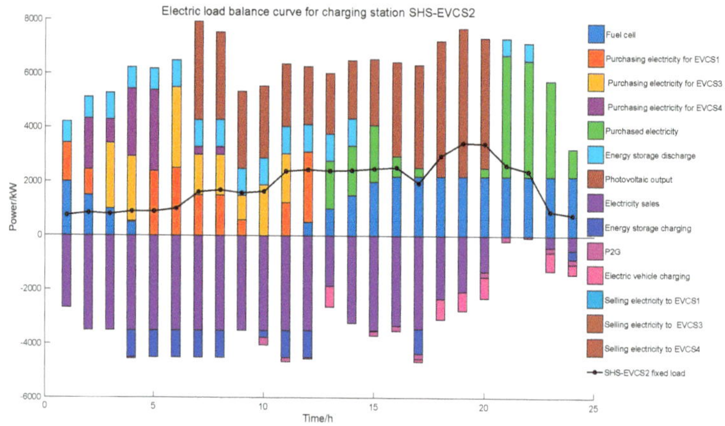

Figure 8. SHS-EVCS2 renewable energy usage and electricity load curve.

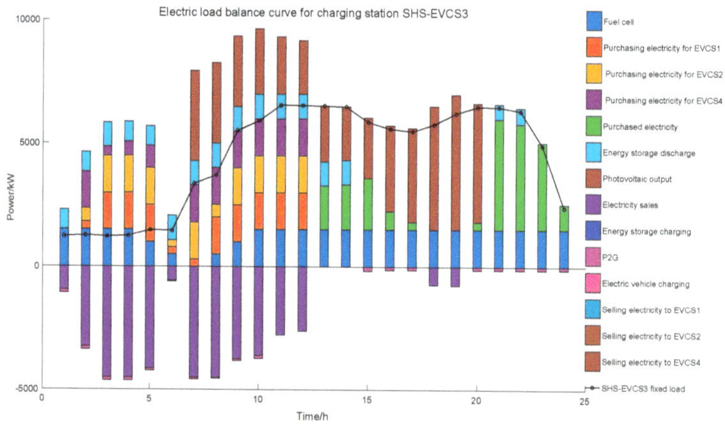

Figure 9. SHS-EVCS3 renewable energy usage and electricity load curve.

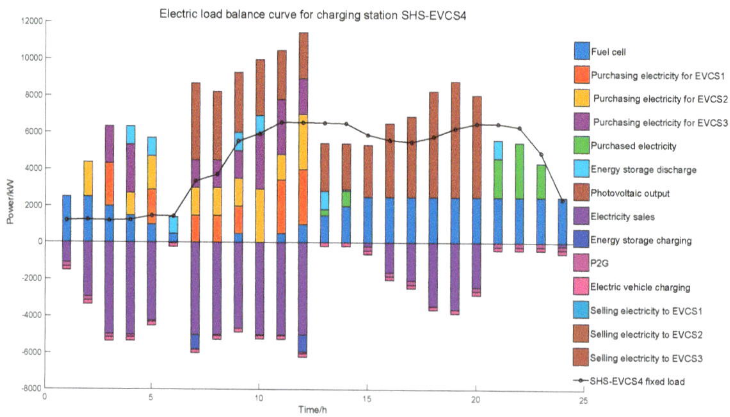

Figure 10. SHS-EVCS4 renewable energy usage and electricity load curve.

Additionally, these infrastructures engage in inter-station power trading, exemplifying the benefits of a decentralized energy network. Optimizing the energy management strategies shown in these figures can lead to the best possible economic benefits while also reducing infrastructure costs. This optimization improves how efficiently each SHS-EVCS operates and makes the power grid more sustainable by using different types of energy storage and generation. Managing energy in this strategic way is key to developing robust, smart power grids and moving toward being able to produce energy independently.

Figure 11 presents a composite graph combining a bar figure and a line graph, illustrating the operational dynamics of an energy storage system over a 24 h period. For the bar figure, positive values represent charging power, indicating the periods when the energy storage system is actively storing energy. Negative values represent discharging power, showing when the system is delivering energy back to four SHS-EVCSs. Notably, during weekends, the charging activities peak around 1 a.m., 6 a.m., 9 a.m., 10 a.m., 1 p.m., 2 p.m., 8 p.m.–10 p.m., possibly correlating with off-peak hours or periods of surplus energy production when travel is less constrained. Discharge events are more sporadic, with significant outputs observed around 7 a.m., 8 a.m., 11 a.m., 12 p.m., 3 p.m.–5 p.m., and 12 a.m., likely aligning with peak demand periods or the operational needs for grid stability. During weekdays, the energy storage undergoes intense charging activities, notably at 4 a.m., 5 a.m., 11 a.m., 12 p.m., 6 p.m., and 7 p.m. The graph highlights these occurrences with blue bars, which correspond to peaks in the battery's state of charge, depicted as red circles on the line graph. This pattern corroborates the anticipated behavior of the battery storage system within the SHS-EVCS framework, efficiently aligning charging and discharging activities with the fluctuations in energy demand and supply dynamics.

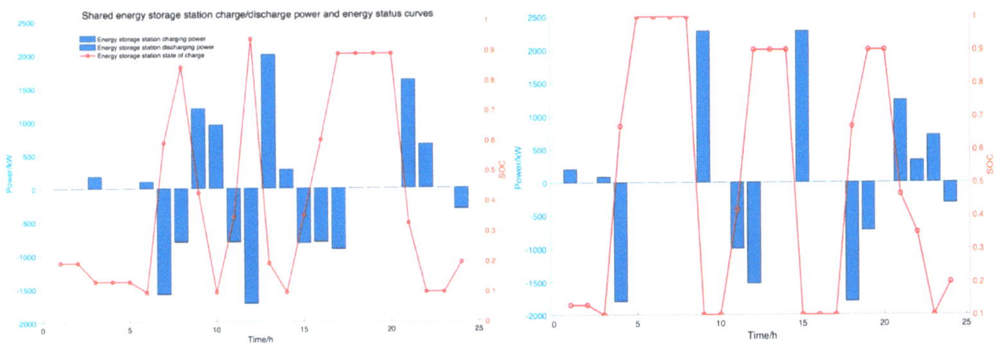

a

b

Figure 11. Battery storage power flow and State of Charge (SOC): (**a**) weekends; (**b**) weekdays.

In the observed data, the state of charge (SOC) of the energy storage system exhibits a cyclic behavior characterized by declines during periods of positive bar representation and increases when the bars are negative. This pattern indicates the regular charging and discharging cycles that the system undergoes. The SOC reaches peak levels at various intervals over a 24 h cycle, aligning with the strategic operational practice of charging during periods of low energy demand and discharging during peak demand periods or when the energy generation is insufficient. Additionally, discharging is implemented when purchasing electricity from the grid or shared battery systems is economically disadvantageous due to higher prices. This cyclical pattern of SOC fluctuations is demonstrative of an effectively managed energy storage system that is dynamically adjusting its charge and discharge cycles. Such management is crucial for leveraging fluctuations in energy prices, enhancing the utilization of generated renewable energy, and thereby increasing the overall stability and efficiency of the SHS-EVCSs. This active modulation not only supports the economic operation within the energy market but also plays a significant role in stabilizing the grid and ensuring a reliable supply of green energy.

Figure 12 shows the diurnal consumption patterns of hydrogen across four SHS-EVCSs. An examination of the hourly data reveals distinct temporal variations in hydrogen utilization. In the early morning hours, from midnight to 10 a.m., CS 1 demonstrates a consistent level of usage, while CS 2 experiences a modest uptick. Conversely, CS 3 and 4 show a decline in usage, with CS 3 encountering a notably sharper fall. After 10 a.m., a significant rise in hydrogen use at CS 3 indicates an increase in operational activities or demand. The remaining charging stations—CS 1, CS 2, and CS 4—also see minor rises in their usage. Until 5 p.m., there's a pronounced increase across all SHS-EVCSs, signaling elevated demand or heightened charging activities. CS 1 registers the largest increase, succeeded by CS 2, CS 4, and CS 3. In the late evening hours, from 10 p.m. to midnight, all stations' activities level off, reaching a steady state of demand. The consumption pattern appears to mirror conventional daily usage trends, characterized by a trough in the early dawn hours and a crest during the evening. These fluctuations are likely reflective of habitual consumer charging behaviors or inherent operational cycles of the SHS-EVCS network.

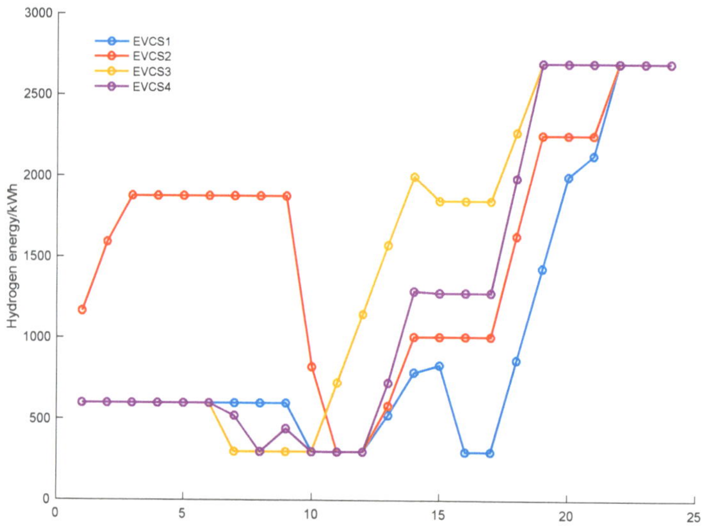

Figure 12. Four SHS-EVCSs' hydrogen usage.

Figure 13 shows an optimization of charging over a 24 h period and how electricity is used throughout the day at four different SHS-EVCSs. The differentiated color within the stacked bar graph distinctly demarcates the electricity usage attributable to each SHS-EVCS, with the proportional segment heights within each hourly bar representing the specific electricity consumption metrics. Upon comparative analysis with Figure 6, which presumably delineates the EV charging quantities at corresponding intervals, a correlative trend emerges. The electricity usage profile depicted in the current figure exhibits parallel fluctuations to the charging activities recorded in Figure 6, suggesting a direct relationship between the electricity consumed by the SHS-EVCSs and the operational charging volume.

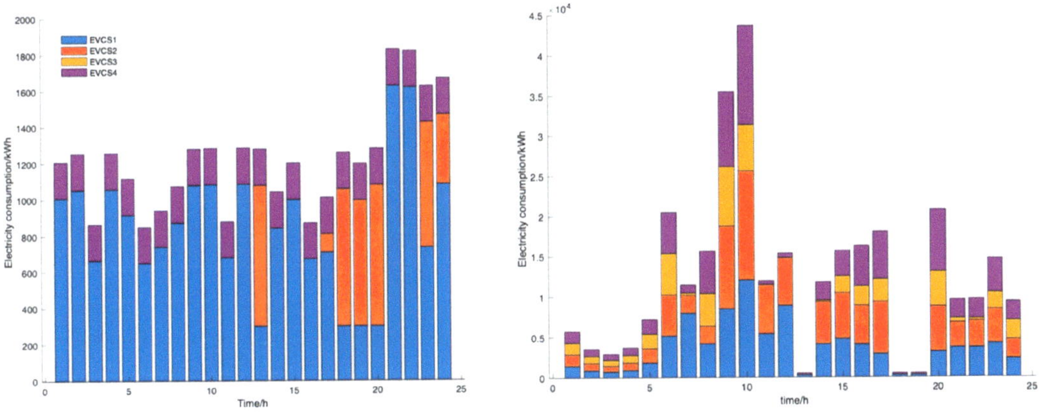

a b.

Figure 13. Four SHS-EVCSs' EV charging optimization over 24 h: (**a**) weekends; (**b**) weekdays.

Each figure in Figure 14 indicates the buyer and seller (for example, "EVCS1 purchases from EVCS2" suggests that source EVCS1 is buying from source EVCS2) with time represented in 24 h. The data points are connected by lines to indicate the change in purchasing activity over time:

(a) EVCS1 purchases from EVCS2: This figure shows the variation in purchases made by source EVCS1 from source EVCS2 over time. There are several distinct peaks in purchasing activity, especially around 3 a.m. and 6 a.m.
(b) EVCS1 purchases from EVCS3: In this figure, purchasing activity by source EVCS1 from source EVCS3 also shows several peaks, particularly at 2 a.m., 4 a.m., 6 a.m., and 9 a.m.
(c) EVCS1 purchases from EVCS4: This figure shows peaks in purchases by source EVCS1 from source EVCS4 at 1 a.m., 5 a.m., 8 a.m., and 12 p.m.
(d) EVCS2 purchases from EVCS1: EVCS2's purchases from source EVCS1 show peaks from 2 a.m. to 12 p.m.
(e) EVCS2 purchases from EVCS3: This figure displays several peaks in purchases by source EVCS2 from source EVCS3, especially at 6 a.m.
(f) EVCS2 purchases from EVCS4: In this figure, source EVCS2's purchases from source EVCS4 peak at 5 a.m.
(g) EVCS3 purchases from EVCS1: Purchases by source EVCS3 from source EVCS1 are higher between 2 a.m. and 12 p.m.
(h) EVCS3 purchases from EVCS2: In this figure, source EVCS3's purchases from source EVCS2 show notable peaks at 2 a.m. and 5 a.m.
(i) EVCS3 purchases from EVCS4: This figure indicates that source EVCS3's purchases from source EVCS4 peak before 12 p.m.
(j) EVCS4 purchases from EVCS1: Source EVCS4's purchases from source EVCS1 show peaks at 12 p.m.
(k) EVCS4 purchases from EVCS2: In this figure, source EVCS4's purchases from source EVCS2 peak at 10 a.m. and 12 a.m.
(l) EVCS4 purchases from EVCS3: Source EVCS4's purchases from source EVCS3 have peaks at 4 a.m., 10 a.m. and 11 a.m.

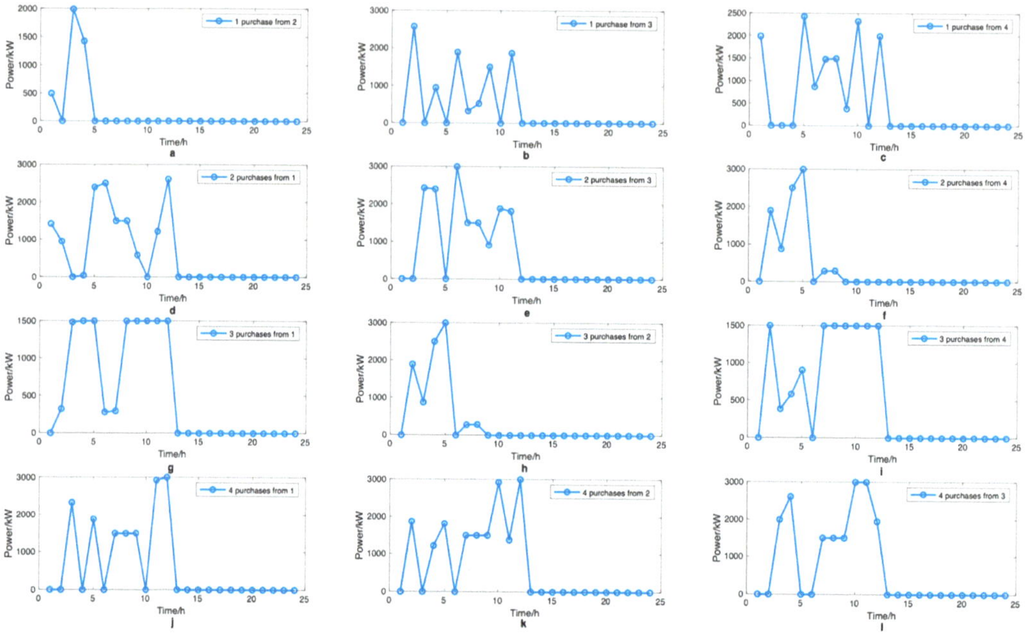

Figure 14. P2P trading strategy between 4 SHS-EVCSs. (**a**) EVCS1 purchases from EVCS2, (**b**) EVCS1 purchases from EVCS3, (**c**) EVCS1 purchases from EVCS4, (**d**) EVCS2 purchases from EVCS1, (**e**) EVCS2 purchases from EVCS3, (**f**) EVCS2 purchases from EVCS4, (**g**) EVCS3 purchases from EVCS1, (**h**) EVCS3 purchases from EVCS2, (**i**) EVCS3 purchases from EVCS4, (**j**) EVCS4 purchases from EVCS1, (**k**) EVCS4 purchases from EVCS2, and (**l**) EVCS4 purchases from EVCS3.

These figures represent the fluctuation of transaction volumes between different suppliers or products over time. Peaks can indicate high demand or bulk transactions at specific points in time. Analyzing these figures could provide insights into the patterns and trends of trade activity between different sources, which may be valuable for optimizing inventory management, forecasting future demands, or adjusting supply chain strategies.

Figure 15 shows the cost comparison between internal electricity purchasing and grid procurement for SHS-EVCSs. The figure demonstrates that the price of internal energy trading daily is more economical than solely relying on electricity purchased from the grid. This comparison highlights the financial advantages of internal trading within SHS-EVCSs, underlining the potential for significant cost savings by optimizing internal electricity transactions over external grid procurement. It provides a clear visual representation of the cost-effectiveness of leveraging internal energy resources, encouraging the adoption of such strategies to enhance economic efficiency in SHS-EVCS operations.

The economic analysis presented here is critical not only for understanding the economic feasibility of SHS-EVCS configurations but also for decision-making related to the scaling and expansion of such systems. The notable disparity between the total daily costs and revenues highlights the potential for considerable profit generation, which can be instrumental in driving investment decisions and fostering the broader adoption of renewable energy technologies within the electric vehicle charging infrastructure.

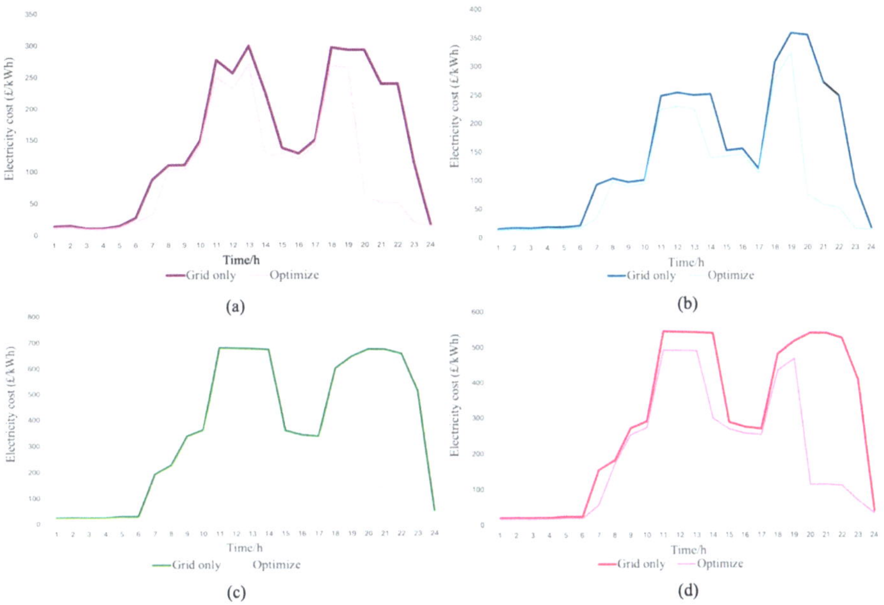

Figure 15. Comparison of internal electricity purchasing versus grid procurement for SHS-EVCSs. (**a**) EVCS1, (**b**) EVCS2, (**c**) EVCS3, and (**d**) EVCS4.

6. Conclusions

This paper introduces a P2P transactive energy trading strategy tailored for multi-SHS-EVCSs, with a particular focus on addressing the challenges posed by the inherent intermittency and volatility of RE generation. This innovative strategy is designed to mitigate the uncertainties arising from inaccurate renewable energy forecasts, which can significantly impact the operational efficiency and economic viability of SHS-EVCSs. A notable advancement presented in this paper is the P2P based cooperative game-theoretical approaches. This agreement acts as a mechanism to resolve conflicts of interest, ensuring that all participating SHS-EVCSs collaborate towards mutual benefit. Such collaboration is crucial for stabilizing the system and preventing any single SHS-EVCS from altering its game strategy—such as electricity pricing strategies—in a manner that could destabilize the system. Overall, the proposed energy trading strategy not only enhances the operational efficiency of SHS-EVCSs but also fosters a cooperative environment that ensures the long-term sustainability and economic efficiency of renewable energy utilization in electric vehicle charging infrastructures.

This study has several limitations that warrant further exploration. Primarily, it does not incorporate demand-side management strategies, notably demand response, into its framework. Moreover, the research overlooks critical distinctions between cooperative and non-cooperative game-theoretical approaches, which could influence the outcomes. Additionally, aspects of social welfare, such as EV driver's welfare, have not been adequately considered. Consequently, future research will need to address these gaps to refine the robustness and applicability of the findings in a broader context.

Author Contributions: Conceptualization, L.D. and C.S.L.; methodology, L.D., Y.Y. and C.S.L.; software, L.D.; validation, L.D.; formal analysis, L.D. and Y.Y.; investigation, L.D.; resources, L.D.; data curation, L.D.; writing—original draft preparation, L.D.; writing—review and editing, L.D., Y.Y., C.S.L. and G.T.; visualization, L.D.; supervision, C.S.L. and G.T.; project administration, C.S.L. and G.T. All authors have read and agreed to the published version of the manuscript.

Funding: This research received no external funding.

Data Availability Statement: Data are contained within the article.

Conflicts of Interest: The authors declare no conflicts of interest.

Nomenclature

$E_{AE,t}$	Hydrogen produced by the electrolyzer, kWh
$P_{AE.t}$	Electrical power consumed by the electrolyzer, kW
v	Hydrogen conversion efficiency, 75%
a	Conversion coefficient for electric energy to the equivalent energy in hydrogen, 39.65 kWh/kg
P_{FC}	Output power of the fuel cell, kW
E_{FC}	Hydrogen consumed by the fuel cell, kWh
η_{FC}s	Energy conversion coefficient for fuel cell, 70%
$E_{hs}(t)$	Total quantity of hydrogen stored in the hydrogen tank at time t, kWh
$E_{hs}(t-1)$	Total quantity of hydrogen stored in the hydrogen tank at time $t-1$, kWh
δ_{hs}	Energy storage decay rate of the hydrogen tank, 5%
$E_{hs}^{in}(t)$	Input quantity of hydrogen into the hydrogen tank, kWh
$E_{hs}^{out}(t)$	Output quantity of hydrogen from the hydrogen tank, kWh
η_{hs}^{in}	Input efficiency of the hydrogen tank, 98%
η_{hs}^{out}	Output efficiency of the hydrogen tank, 98%
P_{STC}	Rated output power of the PV components under standard test conditions
G_{AC}	Real-time solar irradiance intensity
T_c	Real-time temperature of the photovoltaic component surface
G_{STC}	Illumination under standard test conditions, 1000 W/m^2
T_r	25 °C
T_a	Ambient temperature
T_N	*PV* cell operating temperature
$E_{BSS}(c)$	Energy of the battery at the c-th time, kWh
$E_{gen}(c)$	Energy generated by the generator at the c-th time
$E_{load}(c)$	Energy of the load at the c-th time
e_{inv}	Efficiency of the inverter, 95%
η_b^{ch}	Charging efficiency of the battery, 80%
η_b^{dch}	Discharging efficiency of the battery, 80%
$C_{n,t}^{E_H}$	Hydrogen cost, GBP
$C_{n,t}^{FC}$	Fuel cell generator cost, GBP
$C_{n,t}^{grid}$	Grid cost, GBP
$C_{n,t}^{Pv}$	Solar cost, GBP
$C_{n,t}^{p2p}$	P2P trading cost, GBP
C_t^{Bes}	Battery energy storage cost, GBP
$C_{n,t}^{sell}$	Sale revenue, GBP
$\rho_{n,t}^{B}$	P2P trading price by SHS-EVCS n from grid at time t, GBP/kWh
$E_{n,t}^{B}$	P2P trading power by SHS-EVCS n from grid at time t, kWh
$\rho_{n,t}^{g}$	P2P trading price by grid from SHS-EVCS at time t, GBP/kWh
$E_{n,t}^{g}$	P2P trading power by grid from SHS-EVCS at time t, kWh
$\rho_{n,t}^{inter}$	The price of number n SHS-EVCS selling electricity to other SHS-EVCS using internal load, GBP/kWh
$E_{n,t}^{g}$	SHS-EVCSs buying electricity from grid, GBP/kWh
$E_{n,t}^{g*}$	SHS-EVCSs selling electricity to grid, GBP/kWh
$E_{in,t}^{cs}$	SHS-EVCS i follow the SHS-EVCS n pricing rule to buying power at time t, kWh
$\rho_{n,t}^{cs}$	Pricing rule, GBP/kWh

$E_{n,t}^{load}$	Internal load demand of SHS-EVCS, kWh
$E_{n,t}^{i}$	P2P trading power between SHS-EVCS n and I, kWh
$E_{i,t}^{n}$	P2P trading power between SHS-EVCS i and n, kWh
$u_{hs}(t)$	State variable of the electrolyzer
$E_{hs,min}^{o}$	Lower output limit of the electrolyzer, kWh
$E_{hs,max}^{o}$	Upper output limit of the electrolyzer, kWh
$E_{hs,min}^{in}$, $E_{hs,max}^{in}$	Lower and upper input limit of the electrolyzer, kWh
$E_{hs,min}^{out}$, $E_{hs,max}^{out}$	Lower and upper output limit of the electrolyzer, kWh
u_{hs}^{in}	Hydrogen tank storage state variable
u_{hs}^{out}	Hydrogen tank release state variable.
$P_{FC,max}^{out}$	Maximum power output for fuel cell generator, kW
$P_{pv,\ min}$, $P_{pv,\ max}$	Minimum and maximum PV power output, kW
E_{rate_BSS}	Self-discharge rate of the battery
DOD	Depth of discharge, 0.9
Abbreviations	
EV	Electric vehicle
CSs	Charging stations
SHS-EVCSs	Solar-Hydrogen-Battery Storage Electric Vehicle Charging Stations
P2P	Peer to peer
PV	Photovoltaic
RE	Renewable energy
SoC	State of charge
DER	Distributed energy resource
DRP	Demand response program
MILP	Mixed integer linear programming.
DOD	Depth of discharge
P2G	Power to gas
G2P	Gas to power

References

1. Sun, B. A multi-objective optimization model for fast electric vehicle charging stations with wind, PV power and energy storage. *J. Clean. Prod.* **2021**, *288*, 125564. [CrossRef]
2. Merhy, G.; Nait-Sidi-Moh, A.; Moubayed, N. A multi-objective optimization of electric vehicles energy flows: The charging process. *Ann. Oper. Res.* **2021**, *296*, 315–333. [CrossRef]
3. Shahalami, S.; Ahmadnia, S.; Armanfard, S. Locating electric vehicle solar charging and discharging stations using multi-objective genetic algorithm and fuzzy decision making. *Appl. Sol. Energy* **2022**, *58*, 177–186. [CrossRef]
4. Bouguerra, S.; Bhar Layeb, S. Determining optimal deployment of electric vehicles charging stations: Case of Tunis City, Tunisia. *Case Stud. Transp. Policy* **2019**, *7*, 628–642. [CrossRef]
5. Awasthi, A.; Venkitusamy, K.; Padmanaban, S.; Selvamuthukumaran, R.; Blaabjerg, F.; Singh, A. Optimal planning of electric vehicle charging station at the distribution system using hybrid optimization algorithm. *Energy* **2017**, *133*, 70–78. [CrossRef]
6. Mera, L. Application of the Montecarlo simulation for the implementation analysis of fast charging stations for electric vehicles in the Galapagos Province. *Rev. Energ. CENACE* **2021**, *17*, 143–152.
7. Liu, Z.; Song, J.; Kubal, J.; Susarla, N.; Knehr, K.W.; Islam, E.; Nelson, P.; Ahmed, S. Comparing total cost of ownership of battery electric vehicles and internal combustion engine vehicles. *Energy Policy* **2021**, *158*, 112564. [CrossRef]
8. Muratori, M.; Alexander, M.; Arent, D.; Bazilian, M.; Cazzola, P.; Dede, E.M.; Farrell, J.; Gearhart, C.; Greene, D.; Jenn, A.; et al. The rise of electric vehicles—2020 status and future expectations. *Prog. Energy* **2021**, *3*, 022002. [CrossRef]
9. Johnson, L.; Marke, A. The Revival of Electric Vehicles: Historical Perspectives and Modern Advances. *J. Mod. Transp.* **2023**, *31*, 120–135.
10. Zhou, Y.; Lund, P.D. Peer-to-peer energy sharing and trading of renewable energy in smart communities–trading pricing models decision-making and agent-based collaboration. *Renew. Energy* **2023**, *207*, 177–193. [CrossRef]
11. Luo, X.; Shi, W.; Jiang, Y.; Liu, Y.; Xia, J. Distributed peer-to-peer energy trading based on game theory in a community microgrid considering ownership complexity of distributed energy resources. *J. Clean. Prod.* **2022**, *351*, 131573. [CrossRef]
12. Kong KG, H.; Lim, J.Y.; Leong, W.D.; Ng WP, Q.; Teng, S.Y.; Sunarso, J.; How, B.S. Fuzzy optimization for peer-to-peer (P2P) multi-period renewable energy trading planning. *J. Clean. Prod.* **2022**, *368*, 133122. [CrossRef]
13. Yan, X.; Song, M.; Cao, J.; Gao, C.; Jing, X.; Xia, S.; Ban, M. Peer-to-Peer transactive energy trading of multiple microgrids considering renewable energy uncertainty. *Int. J. Electr. Power Energy Syst.* **2023**, *152*, 109235. [CrossRef]

14. Chen, B.; Che, Y.; Zhou, Y.; Zhao, S. Day-ahead optimal peer-to-peer energy trading strategy for multi-microgrids based on Nash bargaining game with data-driven chance constraints. *Sustain. Energy Grids Netw.* **2023**, *36*, 101192. [CrossRef]
15. Gunarathna, C.L.; Yang, R.J.; Jayasuriya, S.; Wang, K. Reviewing global peer-to-peer distributed renewable energy trading projects. *Energy Res. Soc. Sci.* **2022**, *89*, 102655. [CrossRef]
16. Liu, W.; Xu, H.; Wang, X.; Zhang, S.; Hu, T. Optimal dispatch strategy of virtual power plants using potential game theory. *Energy Rep.* **2022**, *8*, 1069–1079. [CrossRef]
17. Choobineh, M.; Arabnya, A.; Khodaei, A.; Zheng, H. Game-theoretic peer-to-peer solar energy trading on blockchain-based transaction infrastructure. *Adv. Electr. Eng. Electron. Energy* **2023**, *5*, 100192. [CrossRef]
18. Goudarzi, A.; Li, Y.; Fahad, S.; Xiang, J. A game theory-based interactive demand response for handling dynamic prices in security-constrained electricity markets. *Sustain. Cities Soc.* **2021**, *72*, 103073. [CrossRef]
19. Mousavizadeh, S.; Haghifam, M.-R.; Shariatkhah, M.-H. A linear two-stage method for resiliency analysis in distribution systems considering renewable energy and demand response resources. *Appl. Energy* **2018**, *211*, 443–460. [CrossRef]
20. Chawal, U.; Rosenberger, J.; Chen VC, P.; Lee, W.J.; Wijemanne, M.; Punugu, R.K.; Kulvanitchaiyanunt, A. A design and analysis of computer experiments based mixed integer linear programming approach for optimizing a system of electric vehicle charging stations. *Expert Syst. Appl.* **2024**, *245*, 123064. [CrossRef]
21. Adouani, Y.; Masmoudi, M.; Jarray, F.; Jarboui, B. Iterative integer linear programming-based heuristic for solving the multiple-choice knapsack problem with setups. *Expert Syst. Appl.* **2024**, *242*, 122835. [CrossRef]
22. Selvam, S.K.; Rajendran, C.; Sankaralingam, G. A linear programming-based bi-objective optimization for forecasting short univariate time series. *Decis. Anal. J.* **2024**, *10*, 100400. [CrossRef]
23. Teng, F.; Zhang, Q.; Xiao, G.; Ban, Z.; Liang, Y.; Guan, Y. Energy Management for a Port Integrated Energy System Based on Distributed Dual Decomposition Mixed Integer Linear Programming. *J. Mar. Sci. Eng.* **2023**, *11*, 1137. [CrossRef]
24. Lee, W.; Han, D.; Won, D. Grid-Oriented Coordination Strategy of Prosumers Using Game-theoretic Peer-to-Peer Trading Framework in Energy Community. *Appl. Energy* **2022**, *326*, 119980. [CrossRef]
25. Yu, X.; Pan, D.; Zhou, Y. A Stackelberg game-based peer-to-peer energy trading market with energy management and pricing mechanism: A case study in Guangzhou. *Sol. Energy* **2024**, *270*, 112388. [CrossRef]
26. Li, Z.; Wu, L.; Xu, Y.; Wang, L.; Yang, N. Distributed tri-layer risk-averse stochastic game approach for energy trading among multi-energy microgrids. *Appl. Energy* **2023**, *331*, 120282. [CrossRef]
27. Paudel, A.; Chaudhari, K.; Long, C.; Gooi, H. Peer-to-Peer Energy Trading in a Prosumer-Based Community Microgrid: A Game-Theoretic Model. *IEEE Trans. Ind. Electron.* **2019**, *66*, 6087–6097. [CrossRef]
28. Malik, S.; Duffy, M.; Thakur, S.; Hayes, B.; Breslin, J. A priority-based approach for peer-to-peer energy trading using cooperative game theory in local energy community. *Int. J. Electr. Power Energy Syst.* **2022**, *137*, 107865. [CrossRef]
29. Li, Z.; Lei, X.; Shang, Y.; Jia, Y.; Jian, L. A Vehicle-to-Vehicle Energy Trading System for Electric Vehicle Networks: A Non-Cooperative Game Approach. *J. Clean. Prod.* **2015**, *414*, 567–589.
30. Liu, Z.; Wu, Q.; Huang, S.; Wang, L.; Shahidehpour, M.; Xue, Y. Optimal Day-Ahead Charging Scheduling of Electric Vehicles Through an Aggregative Game Model. *IEEE Trans. Smart Grid* **2018**, *9*, 5173–5184. [CrossRef]
31. Spiliopoulos, N.; Sarantakos, I.; Nikkhah, S.; Gkizas, G.; Giaouris, D.; Taylor, P.; Rajarathnam, U.; Wade, N. Peer-to-peer energy trading for improving economic and resilient operation of microgrids. *Renew. Energy* **2022**, *199*, 517–535. [CrossRef]
32. Liu, W.; Gu, W.; Wang, J.; Yu, W.; Xi, X. Game Theoretic Non-Cooperative Distributed Coordination Control for Multi-Microgrids. *IEEE Trans. Smart Grid* **2018**, *9*, 6986–6997. [CrossRef]
33. Jing, R.; Xie, M.; Wang, F.; Chen, L. Fair P2P energy trading between residential and commercial multi-energy systems enabling integrated demand-side management. *Appl. Energy* **2020**, *262*, 114551. [CrossRef]
34. Bhatti, B.; Broadwater, R. Energy trading in the distribution system using a non-model based game theoretic approach. *Appl Energy* **2019**, *253*, 113532. [CrossRef]
35. Duan, L.; Guo, Z.; Taylor, G.; Lai, C.S. Multi-Objective Optimization for Solar-Hydrogen-Battery-Integrated Electric Vehicle Charging Stations with Energy Exchange. *Electronics* **2023**, *12*, 4149. [CrossRef]
36. Sapountzi, F.; Gracia, J.; Weststrate, C.; Kee, J.; Fredriksson, H.; Niemantsverdriet, J. Electrocatalysts for the generation of hydrogen, oxygen and synthesis gas. *Prog. Energy Combust. Sci.* **2017**, *58*, 1e35. [CrossRef]
37. Li, N.; Zhao, X.; Shi, X.; Pei, Z.; Mu, H.; Taghizadeh-Hesary, F. Integrated energy systems with CCHP and hydrogen supply: A new outlet for curtailed wind power. *Appl. Energy* **2021**, *303*, 117619. [CrossRef]
38. Ishaq, H.; Dincer, I.; Crawford, C. A review on hydrogen production and utilization: Challenges and opportunities. *Int. J. Hydrogen Energy* **2022**, *47*, 26238e64. [CrossRef]
39. Taleb, B.; Jahjah, R.; Cornu, D.; Bechelany, M.; Al Ajami, M.; Kataya, G.; Hijazi, A.; El-Dakdouki, M.H. Exploring Hydrogen Sources in Catalytic Transfer Hydrogenation: A Review of Unsaturated Compound Reduction. *Molecules* **2023**, *28*, 7541. [CrossRef]
40. Abdelshafy, A.M.; Hassan, H.; Jurasz, J. Optimal design of a grid-connected desalination plant powered by renewable energy resources using a hybrid PSO–GWO approach. *Energy Convers. Manag.* **2018**, *173*, 331–347. [CrossRef]
41. Churkin, A.; Bialek, J.; Pozo, D.; Sauma, E.; Korgin, N. Review of Cooperative Game Theory applications in power system expansion planning. *Renew. Sustain. Energy Rev.* **2021**, *145*, 111056. [CrossRef]
42. China Solar Panel Costs Drop 42% from Year Ago—Report. Reuters. Available online: https://www.reuters.com/world/china/china-solar-panel-costs-drop-42-year-ago-report-2023-12-14/ (accessed on 14 December 2023).

43. Battery Prices Collapsing, Grid-Tied Energy Storage Expanding. PV Magazine. Available online: https://www.pv-magazine.com/2024/03/07/battery-prices-collapsing-grid-tied-energy-storage-expanding/#:~:text=Since%20last%20summer,%20lithium%20battery,/kWh%20to%20$130/kWh (accessed on 7 March 2024).
44. Yin, I. China Targets to Cut Battery Storage Costs by 30% by 2025. S&P Global Commodity Insights. Available online: https://www.spglobal.com/commodityinsights/en/market-insights/latest-news/energy-transition/032222-china-targets-to-cut-battery-storage-costs-by-30-by-2025#:~:text=China's%20electrochemical%20energy%20storage%20cost,the%20Chinese%20Academy%20of%20Sciences (accessed on 22 March 2022).

electronics

Article

Coordinated Reconfiguration with Energy Storage System for Load Restoration in Integrated Electric and Heating Systems

Ke Wang [1,2], Jing Wang [3,*], Pengfei Su [2] and Song Zhang [4]

1 College of Electrical and Power Engineering, Taiyuan University of Technology, Taiyuan 030024, China; wangk50@cardiff.ac.uk
2 School of Engineering, Cardiff University, Cardiff CF24 3AA, UK; sup2@cardiff.ac.uk
3 Hebei Power Exchange Center, Shijiazhuang 050023, China
4 School of Electrical Engineering, Northeast Electric Power University, Jilin 132012, China; zhangsong@neepu.edu.cn
* Correspondence: wjhz0621@126.com

Abstract: Coordinated load restoration of integrated electric and heating systems (IEHSs) has become indispensable following natural disasters due to the increasingly relevant integration between power distribution systems (PDS) and district heating systems (DHS). In this paper, a coordinated reconfiguration with an energy storage system is introduced to optimize load restoration in the aftermath of natural catastrophes. By modifying the DHS network topology, it is possible to maintain an uninterrupted energy supply in unfaulty zones by shifting heat loads among sources and adjusting the operation of coupled devices. Additionally, energy storage systems with rapid response times are implemented to enhance load restoration efficiency, especially when working in conjunction with multiple energy sources. Comprehensive case analyses have been systematically conducted to demonstrate the impact of coordinated reconfiguration with energy storage systems on improving load restoration.

Keywords: district heating system reconfiguration; energy storage system; integrated electric and heating system; load restoration

Citation: Wang, K.; Wang, J.; Su, P.; Zhang, S. Coordinated Reconfiguration with Energy Storage System for Load Restoration in Integrated Electric and Heating Systems. *Electronics* **2024**, *13*, 1931. https://doi.org/10.3390/electronics13101931

Academic Editor: Nikolay Hinov

Received: 1 April 2024
Revised: 8 May 2024
Accepted: 9 May 2024
Published: 15 May 2024

1. Introduction

In the last few years, frequent natural catastrophes have significantly impaired vast energy infrastructures, leading to widespread energy disruptions [1–3]. In 2021, Winter Storm Uri was a significant meteorological event that affected around 10 million people in Texas through the loss of their natural gas and electricity supply, with an estimated economic loss of up to USD 295 billion [4]. In 2022, Hurricane Ian damaged the power and natural gas transmission infrastructure in Florida, United States, affecting over 1.5 million individuals and resulting in economic losses of up to USD 67 billion [5].

With a growing consciousness of potential threats, the load restoration of integrated energy systems has increasingly come into focus [6,7]. The authors of [8] focused on the coordinated load restoration capabilities of district and regional integrated energy systems, aiming to bolster resilience following catastrophic events. The authors of [9] proposed a model for restoring electric and gas systems that incorporates coordination among subsystems. The authors of [10] proposed a restoration strategy considering the power distribution system reconfiguration and optimizing the deployment of repair crews.

The adoption of combined heat and power (CHP) units has established a closer relationship between electric and heating systems [11,12]. This connection has resulted in intricate interactions between the power distribution system (PDS) and the district heating system (DHS), posing two main issues. Firstly, there is the possibility of problem transmission from one system to the other through the coupled components, causing load shedding. For instance, improper switching operation in PDS might influence the heat outputs of the CHP unit, resulting in the unnecessary shedding of heat loads [13]. Secondly, isolated subsystem operations can limit the potential to fully utilize the flexibility of the coupled components, such as CHP units, during the recovery phase. Thus, a collaborative approach is necessary for load restoration.

The reconfiguration of the DHS is a crucial step in the restoration of load in integrated electric and heating systems (IEHSs), analogous to the reconfiguration of PDSs [14]. DHSs can maintain a continuous heat supply in unaffected areas by shifting heat loads among sources and adjusting the network structure. This flexibility in managing heat distribution ensures that the system remains stable and efficient during disruptions. Furthermore, the DHS can enhance the effectiveness of restoration efforts by modifying the operational output of CHP units in coordination with PDS switch operations. This collaborative approach optimizes the recovery of both heat and power, thereby reducing the duration of outages and enhancing the ability of practitioners to respond to emergencies. Such strategic reconfiguration is of paramount importance in scenarios requiring rapid recovery from disruptions, such as severe weather conditions or other events that simultaneously impact power and heating infrastructures. The integration of heat and power restoration strategies serves to enhance the resilience of the IEHS against a range of disruptions, thereby facilitating a swift recovery and operational continuity.

Energy storage systems (ESSs) are becoming increasingly crucial for enhancing system recovery capabilities, particularly due to their rapid response speeds [15,16]. In the event of a system fault, the swift activation of energy storage systems (ESSs), combined with effective control strategies, plays a crucial role in maintaining the stability of power supply, particularly to critical loads. The instant compensatory capabilities of ESSs are pivotal in handling power disruptions. By seamlessly integrating ESSs with multiple energy sources, they ensure that no interruption occurs in the supply of power to essential services during outages. This integration is particularly critical in enhancing the resilience of integrated energy and heating systems (IEHSs) against natural disasters. ESSs facilitate a robust and flexible energy infrastructure that can quickly adapt to sudden changes in power availability. During disruptions caused by severe natural events, such as hurricanes or earthquakes, ESSs provide a reliable and continuous energy supply. This capability is essential for maintaining the operation of vital services like emergency response systems and other critical infrastructure. Moreover, the versatility of ESSs in integrating diverse energy sources enables a smooth transition between these sources during disruptions. This flexibility ensures that essential services remain operational, thereby mitigating the impact of the disaster on the community and economy. The strategic deployment of ESSs within IEHS frameworks significantly bolsters a system's ability to withstand and quickly recover from the catastrophic effects of natural disasters, providing a continuous and reliable energy supply during the most critical times.

This research puts forth a cooperative service-restoration model that incorporates the coordinated reconfiguration of subsystems and ESSs. The substantial contributions are summarized here, as follows:

(1) A model aiming to enable cooperative service restoration is introduced in this paper, accounting for the intricate interplay between the fault isolation and recovery phases. It highlights the fault propagation between the PDSs and DHSs and the collaborative recovery capability of subsystems in resisting natural disasters.

(2) Coordinated reconfiguration is provided to explore the flexibility of time-varying network topologies for service restoration. This strategy enhances the load restoration level by redistributing loads among different energy sources and fine-tuning the energy output of CHP units for improved energy provision.

(3) An energy storage system with fast response speeds is considered in restoration. which can ensure the continuation of power to vital loads. More importantly, the integration of energy storage systems with multiple energy sources can swiftly reinstate significant loads.

Section 2 provides a detailed introduction to the concept of IEHSs. Section 3 unveils a comprehensive model for fault recovery, highlighting the fault propagation and the importance of synchronized reconfiguration with ESS. The results on the P33H13 system and the study for future exploration are revealed in Section 4.

2. A Multi-Time Restoration Model

As shown in Figure 1, the power distribution system and the district heating system are synergistically linked in the IEHS, bridged by CHP units. As the main energy source for both PDSs and DHSs, CHP units significantly strengthen the interconnectivity and mutual support between these subsystems.

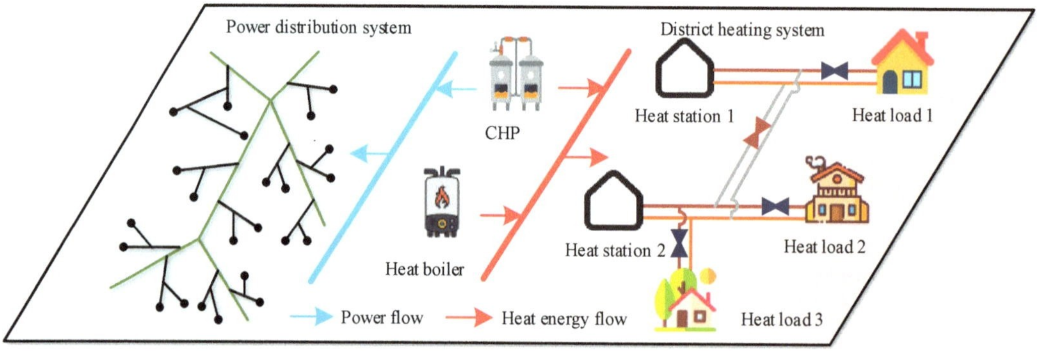

Figure 1. Structure of integrated electric and heating system.

As shown in Figure 2, an innovative collaborative model is specifically designed for service restoration, which integrates both fault isolation and subsequent restoration phases effectively. This model is especially pertinent when dealing with disruptions in complex energy systems like DHS and PDS.

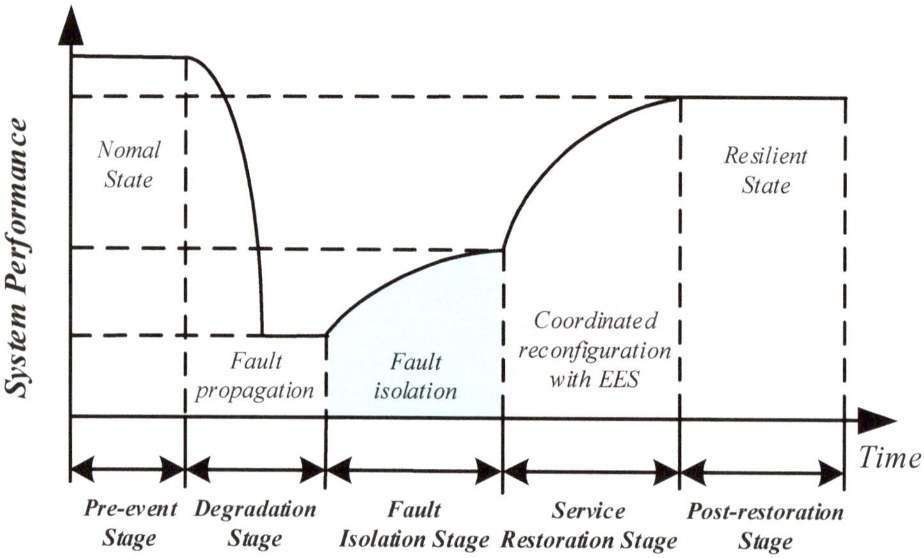

Figure 2. Resilience curve of an IEHS after disaster.

The initial phase, fault isolation, is crucial as it involves the joint operation of the DHS and PDS to pinpoint affected nodes or buses within the network. This operation helps in delineating the boundaries between fault and non-fault areas, which is essential for understanding and managing the scope of fault propagation effectively. By identifying these specific areas, the model facilitates targeted interventions, thereby optimizing the restoration process. Following fault isolation, the recovery phase commences. This phase is characterized by a coordinated reconfiguration of both the DHS and PDS. The objective here is to restore the services to the unaffected regions swiftly and efficiently, thus minimizing the downtime and impact on end users. This reconfiguration is complex, requiring precise control and synchronization between the heating and power sectors to ensure that loads in unaffected areas are restored without introducing new issues.

Moreover, the EES plays a pivotal role in the restoration process. EESs are utilized to maintain the continuity of the energy supply, particularly to vital loads that cannot withstand interruptions. These systems provide a buffer that helps in stabilizing the network during the restoration process by supplying stored energy when it is most needed. The model's effectiveness is further enhanced when integrated with multiple energy sources within an integrated energy and heating system (IEHS). By leveraging various energy sources, the EES can rapidly restore critical loads, thus ensuring that essential services remain operational during most of the restoration period. This integration not only improves the resilience of the system but also enhances its flexibility and responsiveness to disruptions.

2.1. Network Topological Constraints

2.1.1. Fault Isolation Model

In this stage, areas affected by faults show how faults propagate through the subsystem, which could give operators a detailed insight into the dynamics of the network during faults. It accurately delineates fault zones and traces fault propagation through the IEHS, as shown below:

$$(1-f_{ij})(Y_{ij,0}-s_{ij,0}) \le Y_{ij,t} \le (1-f_{ij})Y_{ij,0}, \forall (i,j) \in \mathbb{K}^p \cup \mathbb{K}^l, \forall t \in \mathbb{T}_i, \tag{1}$$

$$\mu_{j,t} - Y_{ij,0} + 1 \ge f_{ij}(1-s_{ij,0}), \forall (i,j) \in \mathbb{K}^p \cup \mathbb{K}^l, \forall t \in \mathbb{T}_i, \tag{2}$$

$$\mu_{j,t} - Y_{ij,0} + 1 \ge f_{ij}(1-s_{ij,0}), \forall (i,j) \in \mathbb{K}^p \cup \mathbb{K}^l, \forall t \in \mathbb{T}_i, \tag{3}$$

$$\mu_{j,t} - Y_{ij,t} + 1 \ge \mu_{j,t}, \forall (i,j) \in \mathbb{K}^p \cup \mathbb{K}^l, \forall t \in \mathbb{T}_i, \tag{4}$$

$$\mu_{j,t} - Y_{ij,t} + 1 \ge \mu_{j,t}, \forall (i,j) \in \mathbb{K}^p \cup \mathbb{K}^l, \forall t \in \mathbb{T}_i, \tag{5}$$

$$\mu_{m,t}^{CHP} = \mu_{s,t}^{CHP}, \forall m \in \mathbb{K}_{i,h}^{CHP}, h \in \mathbb{K}_{i,e}^{CHP}, \forall t \in \mathbb{T}_i, \tag{6}$$

Constraint (1) illustrates that the isolation process can be achieved by the switch/valve in an area that is not affected by faults. Constraints (2)–(3) indicate that if a pipe/line is not equipped with a switch/valve, the occurrence of a fault in it causes its connected nodes/buses to be classified as affected. Constraints (4)–(5) state that nodes/buses of a closed pipe/line are considered to be part of the same area. Constraint (6) specifies that any faults occurring in the CHP units within the DHS or PDS will be mirrored in the other subsystems. These constraints provide insight into the fault isolation dynamics of pipeline networks and establish a framework for prompts and precise fault detection.

2.1.2. Service-Restoration Model

After identifying the fault's exact location, the restoration strategy involves using switches and valves to restore the loads that were not isolated in the previous phase. This restoration procedure follows specific topological constraints.

$$\left(1-f_{ij,c}\right)\left(Y_{ij,t}-s_{ij,0}\right) \leq Y_{ij,t} \leq \left(1-f_{ij,c}\right)\left(Y_{ij,t}+s_{ij,0}\right), \forall(i,j) \in \mathbb{K}^p \cup \mathbb{K}^l, \forall t \in \mathbb{T}_r, \quad (7)$$

$$a_{ij,t}+a_{ji,t}=Y_{ij,t}, \forall(i,j) \in \mathbb{K}^p \cup \mathbb{K}^l, \forall t \in \mathbb{T}_r, \quad (8)$$

$$\sum_{i \in \pi(j)} a_{ij,t} \leq 1, \forall j \in \mathbb{K}^n, \forall t \in \mathbb{T}_r, \quad (9)$$

$$\sum_{s \in \sigma(j)} a_{js,t}=0, \forall j \in \mathbb{K}^n, \forall t \in \mathbb{T}_r, \quad (10)$$

$$Y_{ij,t}=N_{ij}-N_s, \forall(i,j) \in \mathbb{K}^p \cup \mathbb{K}^l, \forall t \in \mathbb{T}_r, \quad (11)$$

$$\mu_{j,t}-Y_{ij,t}+1 \geq \mu_{j,t}, \forall(i,j) \in \mathbb{K}^p \cup \mathbb{K}^l, \forall t \in \mathbb{T}_r, \quad (12)$$

$$\mu_{j,t}-Y_{ij,t}+1 \geq \mu_{j,t}, \forall(i,j) \in \mathbb{K}^p \cup \mathbb{K}^l, \forall t \in \mathbb{T}_r, \quad (13)$$

Constraint (7) illustrates that switches/valves in non-faulted areas are crucial for the network's structural adjustment. Only the switch/valve equipped on the non-faulted pipe/line can be utilized for network reconfiguration. Constraints (8)–(10) illustrate that maintaining a radial network topology is essential in ensuring that energy flow follows a single route [17,18]. Constraints (12)–(13) illustrate that the nodes/buses of a closed pipe/line are considered to be part of the same area.

2.2. *Operation Constraints*

2.2.1. PDS Operation Constraints

A model for collaborative service restoration has been developed using mixed-integer second-order cone programming.

1. Power Balance Constraints

$$P_{j,t}=\sum_{s \in \delta(j)} P_{js,t}-\sum_{i \in \pi(j)}\left(P_{ij,t}-R_{ij} L_{ij,t}\right), \forall j \in \mathbb{K}^b, \forall t \in \mathbb{T}, \quad (14)$$

$$Q_{j,t}=\sum_{s \in \delta(j)} Q_{js,t}-\sum_{i \in \pi(j)}\left(Q_{ij,t}-X_{ij} L_{ij,t}\right), \forall j \in \mathbb{K}^b, \forall t \in \mathbb{T}, \quad (15)$$

$$P_{j,t}=P_{j,t}^{DG}+P_{j,t}^{CHP}+P_{j,t}^{EES}-\left(P_{j,t}^{L}-P_{j,t}^{Loss}\right), \forall j \in \mathbb{K}^b, \forall t \in \mathbb{T}, \quad (16)$$

$$Q_{j,t}=Q_{j,t}^{DG}+Q_{j,t}^{CHP}+Q_{j,t}^{EES}-\left(Q_{j,t}^{L}-Q_{j,t}^{Loss}\right), \forall j \in \mathbb{K}^b, \forall t \in \mathbb{T}, \quad (17)$$

$$\left\|2P_{ij,t}\ 2Q_{ij,t}\ L_{ij,t}-U_{i,t}\right\|_2 \leq L_{ij,t}+U_{i,t}, \forall j \in \mathbb{K}^b, \forall t \in \mathbb{T}, \quad (18)$$

Constraints (14)–(17) illustrate the power flow at each bus. To improve the efficiency of the solution, the nonconvex current constraint is relaxed to a second-order cone constraint in (18). The treatment has been extensively used and justified in distribution systems [19,20].

2. Transmission Capacity Constraints

$$\left|P_{ij,t}\right| \leq Y_{ij,t} S_{ij}^{Max}, \forall(i,j) \in \mathbb{K}^l, \forall t \in \mathbb{T}, \quad (19)$$

$$\left|Q_{ij,t}\right| \le Y_{ij,t} S_{ij}^{Max}, \forall (i,j) \in \mathbb{K}^l, \forall t \in \mathbb{T}, \tag{20}$$

Constraints (19)–(20) illustrate that the transmission power of a closed line should be within the bounds. The transmission power of an open line should be 0.

3. Voltage Drop Constraints

$$-(1 - Y_{ij,t})M \le U_{i,t} - U_{j,t} - 2(R_{ij}P_{ij,t} + X_{ij}Q_{ij,t}) + \left(R_{ij}^2 + X_{ij}^2\right)L_{ij,t} \le (1 - Y_{ij,t})M, \quad \forall (i,j) \in \mathbb{K}^l, \forall t \in \mathbb{T}, \tag{21}$$

$$U_j^{Min} \le U_{j,t} \le U_j^{Max}, \forall j \in \mathbb{K}^b, \forall t \in \mathbb{T}, \tag{22}$$

Constraint (21) illustrates that the voltage drops along a closed line. Constraint (22) illustrates that the voltage variation along the line should be within the bounds.

4. Unit Generation Constraints

$$(1 - \mu_{j,t})P_{CHP,j}^{Min} \le P_{j,t}^{CHP} \le (1 - \mu_{j,t})P_{CHP,j}^{Max}, \forall j \in \mathbb{K}^c, \forall t \in \mathbb{T}, \tag{23}$$

$$(1 - \mu_{j,t})Q_{CHP,j}^{Min} \le Q_{j,t}^{CHP} \le (1 - \mu_{j,t})Q_{CHP,j}^{Max}, \forall j \in \mathbb{K}^c, \forall t \in \mathbb{T}, \tag{24}$$

$$(1 - \mu_{j,t})P_{DG,j}^{Min} \le P_{j,t}^{DG} \le (1 - \mu_{j,t})P_{DG,j}^{Max}, \forall j \in \mathbb{K}^d, \forall t \in \mathbb{T}, \tag{25}$$

$$(1 - \mu_{j,t})Q_{DG,j}^{Min} \le Q_{j,t}^{DG} \le (1 - \mu_{j,t})Q_{DG,j}^{Max}, \forall j \in \mathbb{K}^d, \forall t \in \mathbb{T}, \tag{26}$$

Constraints (23)–(26) illustrate that the power generation of CHP units and DG should be within the bounds. When the power sources are cut off, the power generation is reduced to 0.

5. Load Shedding Constraints

$$\mu_{j,t}P_j^L \le P_{j,t}^{Loss} \le P_j^L, \forall j \in \mathbb{K}^d, \forall t \in \mathbb{T}, \tag{27}$$

$$\mu_{j,t}Q_j^L \le Q_{j,t}^{Loss} \le Q_j^L, \forall j \in \mathbb{K}^d, \forall t \in \mathbb{T}, \tag{28}$$

Constraints (27)–(28) demonstrate that, in faulted areas leading to unit shutdowns, electric loads will be completely removed, and in non-faulted areas, a portion of the loads will be shed to maintain power balance.

6. ESS Operation Constraints

$$E_{j,t}^{EES} = \begin{cases} E_{j,t-1}^{EES} - \frac{P_{j,t}^{EES} \cdot \Delta t}{\delta_j^{Dis}}, & if\ P_{j,t}^{EES} > 0 \\ E_{j,t-1}^{EES} - P_{j,t}^{EES} \cdot \Delta t \cdot \delta\delta_j^{Ch}, & if\ P_{j,t}^{EES} \le 0 \end{cases} \tag{29}$$

Constraint (29) demonstrates the dynamic limit of state for the energy stored in ESS and the charging/discharging power for two consecutive time slots, taking into account the charging/discharging efficiencies. It is important to specify the power flow direction of the ESS. When the value of $P_{j,t}^{EES}$ is positive, it indicates that the ESS is discharging, and when it is negative, it indicates that the ESS is charging.

$$SoC_{j,t} = \frac{E_{j,t}^{EES}}{E_j^{Cap.}}, \quad \forall j \in \mathbb{K}^e, \forall t \in \mathbb{T}, \tag{30}$$

$$SoC_j^{Min} \leq SoC_{j,t} \leq SoC_j^{Max},\ \forall j \in \mathbb{K}^e, \forall t \in \mathbb{T}, \tag{31}$$

Constraints (30)–(31) define the charge capacity boundaries for the ESS to prevent both excessive charging and discharging.

$$(1-\mu_{j,t})P_{EES,j}^{Min} \leq P_{j,t}^{EES} \leq (1-\mu_{j,t})P_{EES,j}^{Max}, \forall j \in \mathbb{K}^e, \forall t \in \mathbb{T}, \tag{32}$$

Constraint (32) illustrates the capacity limit of EES.

2.2.2. DHS Operation Constraints

In the energy flow model, the quantity of heat quantity is represented by an alternative variable, i.e., $H_{ij} = CM_{ij}\left(\tau_{ij}^S - \tau_{ij}^R\right)$, which is then incorporated into the service-restoration model [21–23].

1. Heat Station Constraints

$$\varsigma_j^{Min} H_{j,t}^{CHP} \leq P_{j,t}^{CHP} \leq \varsigma_j^{Max} H_{j,t}^{CHP}, \forall j \in \mathbb{K}^c, \forall t \in \mathbb{T}, \tag{33}$$

$$H_{j,t}^{HB} = \theta_j F_{j,t}^{HB}, \forall j \in \mathbb{K}^h, \forall t \in \mathbb{T}, \tag{34}$$

$$\sum_{j\in\mathbb{K}_i^c} H_{j,t}^{CHP} + \sum_{j\in\mathbb{K}_i^h} H_{j,t}^{HB} = H_{i,t}^{HS}, \forall i \in \mathbb{K}^{hs}, \forall t \in \mathbb{T}, \tag{35}$$

Constraint (33) illustrates the power-to-heat ratio of the CHP unit. Constraint (34) illustrates the ratio of fuel consumption to the production of a heating boiler (HB). Constraint (35) illustrates that heat generation in the heating station (HS) originates from the CHP unit and HB [24].

2. Heat Transmission Constraints

$$H_{ij,t}^{Out} = H_{ij,t}^{In} - H_{ij,t}^{Loss}, \forall (i,j) \in \mathbb{K}^p, \forall t \in \mathbb{T}, \tag{36}$$

$$\left|H_{ij,t}^{In}\right| \leq Y_{ij,t} H_{ij}^{Max}, \forall (i,j) \in \mathbb{K}^p, \forall t \in \mathbb{T}, \tag{37}$$

$$\left|H_{ij,t}^{Out}\right| \leq Y_{ij,t} H_{ij}^{Max}, \forall (i,j) \in \mathbb{K}^p, \forall t \in \mathbb{T}, \tag{38}$$

Constraint (36) illustrates the heat quantity loss from the inlet to the outlet pipe. Constraints (37)–(38) illustrate that the heat transmission along the pipe should be within the bounds. The heat transmission of an open line should be 0.

3. Energy Balance Constraints

$$\sum_{(j,s)\in S_j^{pipe-}} H_{js,t}^{Out} + \sum_{k\in\mathbb{K}_j^{HS}} H_{k,t}^{HS} = H_{j,t}^L - H_{j,t}^{Loss} + \sum_{(i,j)\in S_j^{pipe+}} H_{ij,t}^{In}, \forall j \in \mathbb{K}^n, \forall t \in \mathbb{T}, \tag{39}$$

Constraint (39) illustrates the heat flow at each node.

4. Unit Generation Constraints

$$(1-\mu_{j,t})H_{CHP,j}^{Min} \leq H_{j,t}^{CHP} \leq (1-\mu_{j,t})H_{CHP,j}^{Max}, \forall j \in \mathbb{K}^c, \forall t \in \mathbb{T}, \tag{40}$$

Constraint (40) illustrates that the heat generation of CHP units and HB should be within the bounds. When the heat sources are cut off, the heat generation is reduced to 0.

5. Load Shedding Constraints

$$\mu_{j,t}H_j^L \leq H_{j,t}^{Loss} \leq H_j^L, \forall j \in \mathbb{K}^n, \forall t \in \mathbb{T}, \tag{41}$$

Constraint (41) demonstrates that in regions affected by faults, heat loads will be completely shed, whereas areas without faults will experience partial load losses.

2.3. Objective and Resilience Metrics

The objective stated in (42) is to minimize the reduction in electrical and thermal loads during the fault restoration process. The resilience metric in (43) is to assess the level of load recovery with the proposed strategy.

$$\min\left\{\Delta T_i\left(\sum_{j\in\mathbb{K}^b}\alpha_j P_{j,t}^{Loss}+\sum_{j\in\mathbb{K}^n}\beta_j H_{j,t}^{Loss}\right)+\Delta T_r\left(\sum_{j\in\mathbb{K}^n}\alpha_j P_{j,t}^{Loss}+\sum_{j\in\mathbb{K}^n}\beta_j H_{j,t}^{Loss}\right)\right\}+P_t^{Line}, \tag{42}$$

$$R_c = 1-\frac{\Delta T_i\left(\sum_{j\in\mathbb{K}^b}\alpha_j P_{j,t}^{Loss}+\sum_{j\in\mathbb{K}^n}\beta_j H_{j,t}^{Loss}\right)+\Delta T_r\left(\sum_{j\in\mathbb{K}^b}\alpha_j P_{j,t}^{Loss}+\sum_{j\in\mathbb{K}^n}\beta_j H_{j,t}^{Loss}\right)}{(\Delta T_i+\Delta T_r)\left(\sum_{j\in\mathbb{K}^b}\alpha_j P_{j,t}^{L}+\sum_{j\in\mathbb{K}^n}\beta_j H_{j,t}^{L}\right)}. \tag{43}$$

3. Case Studies

3.1. Case Description

The proposed methodology is thoroughly tested using a specially adapted version of the P33H13 system, which is detailed in Figure 3. The modified 33-bus PDS is based on a standard IEEE 33-bus case, and the 14-node DHS is designed based on the 14-node DHS in [25] according to the design code of district heating network (CJJ34-2016). This system comprises three distinct heating stations, each equipped with advanced extraction condensing CHP units and a supplemental heating boiler. These installations are designed to efficiently meet the heating demands of the system.

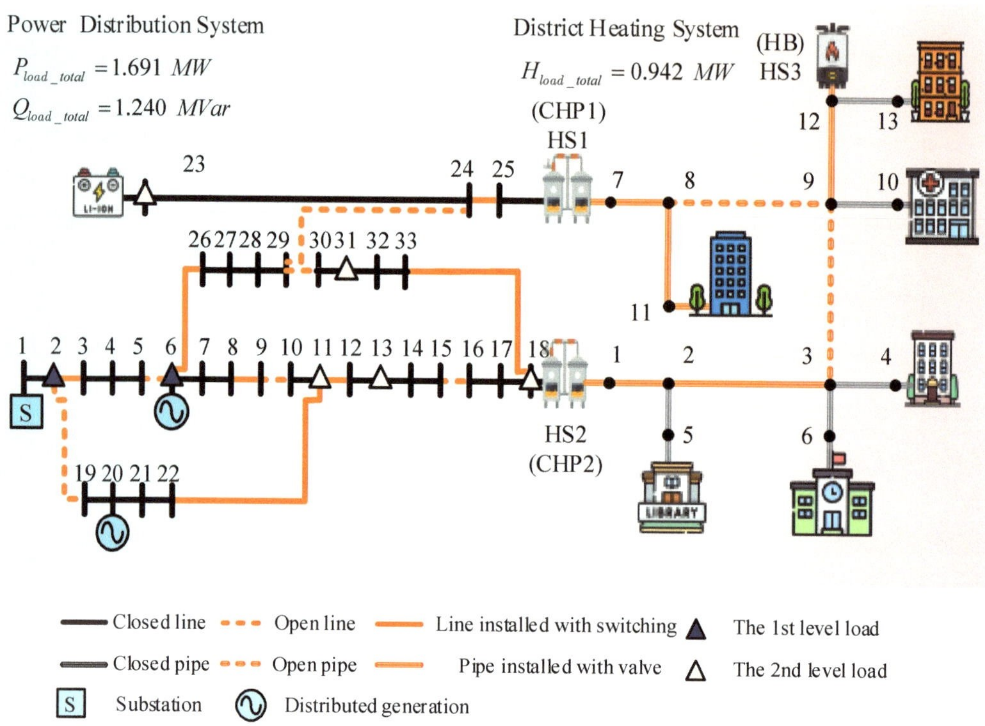

Figure 3. Structure of P33H13 system.

Prior to any significant events, particular operational setups are made—specifically, certain valves and switches are strategically left open. This configuration is critical as it allows for a seamless transition and flexibility in operations during routine functions or in the event of system adjustments. Moreover, the IEHS incorporated within the setup includes small-scale CHP units. These micro-gas turbines play a pivotal role as they link the park-level heating system with the distribution power system, thereby enhancing the system's integration and operational efficiency.

Figure 3 provides a clear visual representation of the connections and layout of these components, emphasizing the synergy between the park-level heating and power distribution systems. This integration is essential for optimizing the energy flow and ensuring stability across both networks. The system's capacity to handle electrical and heat loads was rigorously evaluated. The total electric and reactive power loads of the system are recorded at 1.691 MW and 1.240 MVar, respectively, while the thermal load is approximately 0.942 MW. Both the DG and CHP units possess a capacity of 0.5 MVA and 0.5 MW, respectively, illustrating a well-balanced distribution of power and heating capabilities across the system. The experiments are carried out on a computer equipped with an i7-1165G7 CPU and 16 GB of memory, which is programmed by Matlab R2020a.

3.2. Case Analysis

To illustrate the effect of synchronized reconfiguration with EES, three cases are examined:

Case 1: Load restoration only by network reconfiguration in PDS.

Case 2: Load restoration considering coordinated network reconfiguration.

Case 3: Load restoration incorporating both the coordinated reconfiguration strategy and the utilization of an energy storage system.

3.2.1. PDS Fault Scenario

In this scenario, lines LB6-26, LB8-9, LB11-12, and LB23-24 suffered destruction due to natural catastrophes. This led to significant power and heating disruptions across the IEHS. The effects of these incidents are detailed in Table 1, from which conclusions can be inferred.

Table 1. Load restoration and resilience metrics.

Scenario	Case	Complete Load Restoration (MW)	Load Restoration (MW)		Resilience Metrics
			Electric Load	Heat Load	
PDS	Case 1	29.65	22.33	7.32	0.78
	Case 2	35.79	22.63	13.16	0.85
	Case 3	36.51	23.35	13.16	0.88
DHS	Case 1	25.11	17.85	7.26	0.73
	Case 2	28.85	18.21	10.64	0.76
	Case 3	30.24	19.2	11.04	0.80

The initial interference with line LB23-24 in the PDS had repercussions in the DHS, mediated by the CHP units, which led to restricted output from CHP1 during the fault isolation stage. This resulted in the total loss of the second-level electric load at bus 23 and the partial diminishment of heat loads at nodes 8 and 11.

Secondly, a coordinated reconfiguration is carried out, which involves the remote orchestration of valves/switches and the equitable distribution of loads across multiple energy sources. In Case 2, valves implemented on pipes PN3-9, PN7-8, and PN8-9 redistribute the heat loads of the CHP2 unit, reducing the amount of load shedding within the DHS.

Thirdly, energy storage systems can provide quick backup electricity to ensure the continuous operation of critical facilities and services. In Case 3, the switch implemented on lines LB24-29 transfers the electric loads at buses 23, 24, 26, 27, 28, and 29 to the energy storage system, reducing the amount of load shedding within the PDS.

As shown in Figure 4, as the application of energy storage and grid reconfiguration strategies is implemented, there is a gradual increase in the amount of load recovery observed. The needed time for restoration and fault clearance in Cases 2 and 3 are 10 min and 7 min. These technologies enhance the system's capacity to rapidly adapt and restore power and heat loads following disruptions, thereby ensuring a more resilient and efficient energy supply. The progressive deployment of storage and restructuring measures effectively mitigates the impact of outages, leading to improved stability and reduced restoration time in energy networks.

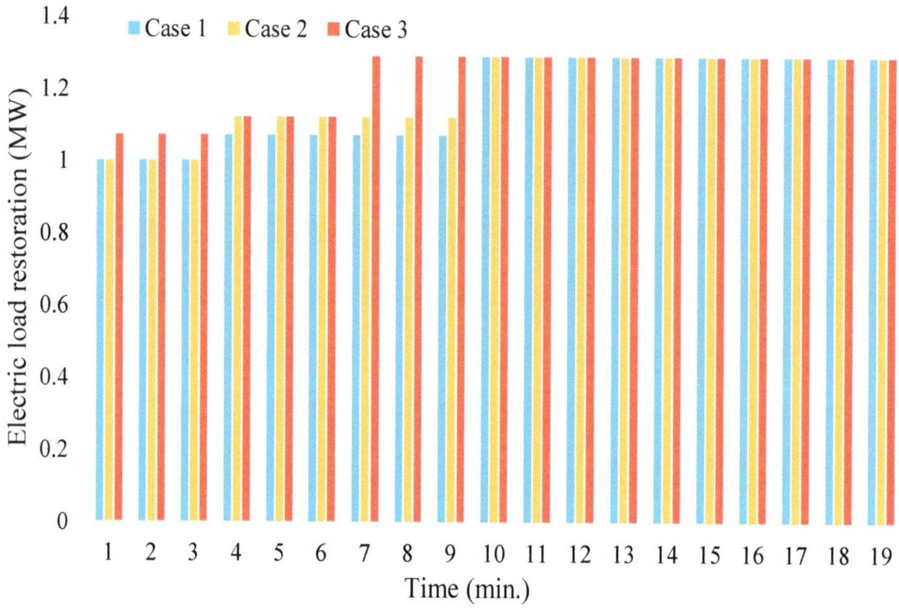

(a)

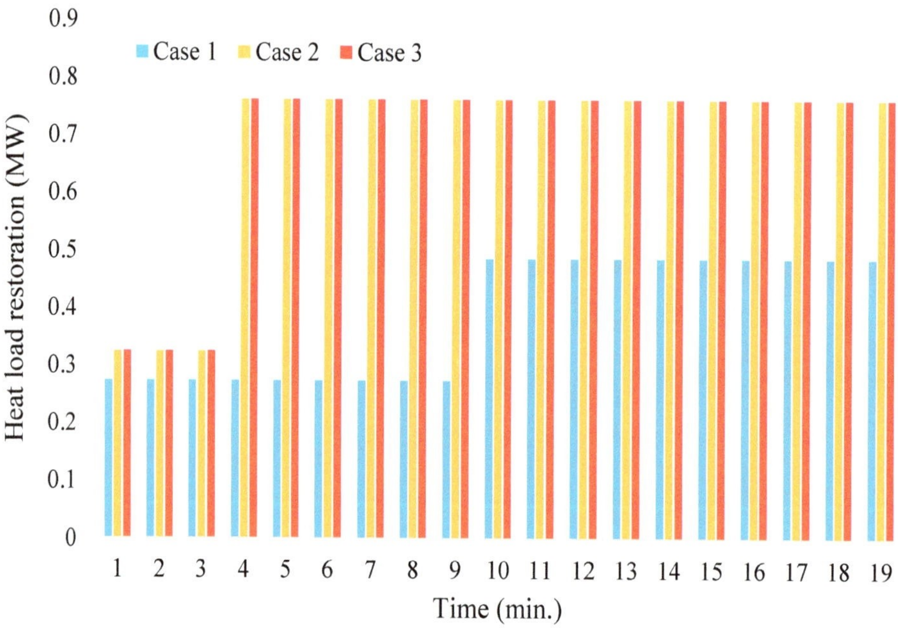

(b)

Figure 4. Load restoration. (**a**) Electric loas restoration. (**b**) Heat loas restoration.

3.2.2. DHS Fault Scenario

Table 1 elucidates the sequence of switch and valve operations after disasters affected pipes PN2-3, PN2-11, and PN12-13. Table 1 provides a comprehensive illustration of the total load recovery amounts under distinct fault scenarios and cases, with a particular focus on the electrical and thermal load shedding quantities. Furthermore, the table includes resilience metrics that quantify the effectiveness of the system in withstanding and recovering from these disturbances. By comparing the resilience indicators across different cases, the table offers a comprehensive overview of the system resilience in maintaining functionality during and after faults. This allows for an in-depth analysis of the effectiveness of implemented measures.

Firstly, the occurrence of a disruption in pipe PN2-3 within the DHS leads to significant interruptions in both power and heat supply. This disruption directly affects the heat loads at nodes 3, 4, and 6, which in turn causes a noticeable decrease in electrical loads at buses 30 and 31. The primary reason for this decrease is the diminished energy production capacity of the CHP2. Such reductions in energy output necessitate immediate adjustments to prevent extensive service downtimes.

To address these challenges and enhance the resilience of the system, adjustments to the configuration of the district heating network are essential. One effective strategy is the redistribution of heat loads, particularly in response to disaster scenarios. For instance, in Case 2, the heat loads at nodes 3, 4, and 6 are efficiently redistributed. This is achieved through strategic valve operations within the network, specifically in pipes PN3-9, PN8-9, and PN9-12. By manipulating the flow through these critical junctions, it is possible to improve overall energy distribution.

Furthermore, such adjustments contribute significantly to the operational flexibility of the CHP1. This flexibility is crucial for the quick restoration of system load, especially in times of unexpected failures or maintenance activities. The enhanced flexibility not only helps in managing the immediate effects of the disruption but also prepares the network for future contingencies by improving its adaptive capabilities.

Thirdly, the implementation of ESS stands out as a key measure to provide rapid backup power and ensure the uninterrupted functioning of essential facilities and services. In Case 3, the operationalization of a switching mechanism along the LB24-29 lines effectively channels the electrical loads of buses 23, 24, 26, 27, 28, and 29 toward an energy storage system. This strategic maneuver significantly reduces load shedding within the PDS.

3.2.3. Discussion

A coordinated reconfiguration strategy with EES is more effective for load restoration in IEHS. Under the PDS fault scenario, there is a significant increase in load restoration (2.0%, 23.1%) and a notable enhancement in the resilience metric (3.5%, 12.8%) in Case 3 compared to Cases 1 and 2. Under the DHS fault scenario, a comparative analysis reveals a significant increase in load restoration (4.8% 20.4%) and a significant improvement in the resilience metric (5.3%, 9.6%) in Case 3 when compared to the results of Cases 1 and 2.

4. Conclusions

In this study, we introduce a strategy for service restoration that effectively combines network reconfiguration and EES within IEHS. This method meticulously considers the dynamic interactions between fault isolation and service restoration phases, elucidating the complex dependency between PDS and DHS. The main conclusions are summarized here, as follows: (i) the disturbances within the PDS significantly influence the DHS operations via interconnected elements; (ii) the strategic adjustments in DHS configurations to enable them to cooperate with PDS reconfiguration can enhance the flexibility of energy distribution, by effectively reallocating loads among various energy sources; (iii) the EES can mitigate the effects of load reductions during cross-system disturbances. By precisely adjusting the output of EES, we have significantly enhanced the resilience of IEHS, quantifiably improving energy efficiency and system reliability under stress conditions. This targeted approach ensures a more robust and responsive infrastructure, leading to an average resilience improvement of 10.32% across tested scenarios.

Moving forward, we aim to delve into a broader spectrum of uncertainties, including variability in demand and unexpected system failures. Fundamentally, our research underscores the critical role of coordinated network reconfiguration in fortifying the robustness of IEHSs and offers valuable directions for forthcoming investigations in this field. Also, a further study for service restoration considering the time scales between PDSs and DHSs, the economic costs of switch/valve operations, and the utilization of energy storage systems will be performed and reported in the future.

Author Contributions: Writing—original draft, K.W.; Writing—review & editing, J.W., P.S. and S.Z. All authors have read and agreed to the published version of the manuscript.

Funding: This research received no external funding.

Data Availability Statement: Data are contained within the article.

Conflicts of Interest: The authors declare no conflict of interest.

Nomenclature

A	*Indices and Sets*
t	Index of time
$\mathbb{T}_i$	Index of fault isolation phase
$\mathbb{T}_r$	Index of load restoration phase
$\mathbb{K}_i^{CHP}$	Set of CHP units installed in heat station i
$\mathbb{K}^p/\mathbb{K}^l$	Set of transmission lines and heating pipes
$\mathbb{K}^b/\mathbb{K}^n$	Set of PDS buses and DHS nodes
$\mathbb{K}^c$	Set of CHP units
$\mathbb{K}^h$	Set of heating boilers
$\mathbb{K}^{hs}$	Set of heating exchange stations
$k_j^{pipe,\text{in}}/k_j^{pipe,out}$	Set of pipes originating from/leading to node j
B	*Parameters*
$Y_{ij,0}$	Boolean variable reflecting the operational/non-operational status of line/pipe (i,j) in normal condition
$s_{ij,0}$	Boolean variable, the valve/switch on pipe/line (i,j) is engaged and open when $s_{ij,0}=1$, the pipe/line does not contain any switch/valve or it is in a closed state when $s_{ij,0}=0$
f_{ij}	Boolean variable that represents if line/pipe (i,j) has incurred damage due to a catastrophic event.
n_s	Quantity of nodes/buses
n_{ij}	Quantity of transmission lines/heating pipes
R_{ij}/X_{ij}	Resistance and reactance of line (i,j)
S_{ij}^{Max}	Maximum transmission capability of line (i,j)
U_j^{Min}/U_j^{Max}	Range of squared voltage magnitudes from minimum to maximum the bus j
$P_{DG,j}^{Min}/P_{DG,j}^{Max}$	Range of power output of distributed generation j
$P_{CHP,j}^{Min}/P_{CHP,j}^{Max}$	Range of power output of CHP unit j
$P_{EES,j}^{Min}/P_{EES,j}^{Max}$	Range of power output of EES j from minimum to maximum
Δt	Time interval
$\delta_j^{Dis}/\delta_j^{Ch}$	Charging/discharging efficiency of ESS j
$E_j^{Cap.}$	Capacity of ESS j

SoC_j^{Min}/SoC_j^{Max}	Range of state of charge of ESS j
$\varsigma_j^{Min}/\varsigma_j^{Max}$	Range of power to heat ratio of CHP unit j from minimum to maximum
θ_j	Proportion of fuel consumption to heat generation of HB j
H_{ij}^{Max}	Transmission capability of heat quantity of pipe (i,j)
α_j/β_j	Weight factor of electric and heat load j
$\Delta T_i/\Delta T_r$	Fault isolation and load restoration duration
C	*Variables*
$Y_{ij,t}$	Boolean variable reflecting the operational/non-operational status of line/pipe (i,j) at t
$\mu_{j,t}$	Boolean variable that illustrates the status of a bus/node i as either functioning or malfunctioning.
$L_{ij,t}$	Current magnitude squared of line (i,j) at t
$P_{j,t}/Q_{j,t}$	Power infusion of bus i at t
$P_{js,t}/Q_{js,t}$	Power flow from bus j to bus s at t
$P_{j,t}^{DG}/Q_{j,t}^{DG}$	Power output of distributed generation j at t
$P_{j,t}^{CHP}/Q_{j,t}^{CHP}$	Power output of CHP unit j at t
$P_{j,t}^{EES}/Q_{j,t}^{EES}$	Charging/discharging power of EES j at t
$E_{j,t}^{EES}$	Energy stored in ESS at t
$P_{j,t}^{L}/Q_{j,t}^{L}$	Electrical load of bus j at t
$P_{j,t}^{Loss}/Q_{j,t}^{Loss}$	Decrease in electrical load of bus j at t
$U_{i,t}$	Squared voltage magnitude of bus j at t
$H_{j,t}^{CHP}$	Heat output of CHP unit j at t
$H_{j,t}^{HB}$	Heat output of heating boiler j at t
$F_{j,t}^{HB}$	Amount of fuel expended by the heating boiler j at t
$H_{k,t}^{HS}$	Heat output of heat station k at t
$H_{ij,t}^{In}/H_{ij,t}^{Out}$	Thermal energy flow into and out of the pipeline (i,j) at t
$H_{ij,t}^{Loss}$	Heat quantity reduction within the pipe (i,j) at t
$H_{j,t}^{Loss}$	Heat quantity reduction of node j at t
P_t^{Line}	Power losses in transmission of line (i,j) at t

References

1. Wang, C.; Wei, W.; Wang, J.; Liu, F.; Qiu, F.; Correa-Posada, C.M.; Mei, S. Robust Defense Strategy for Gas–Electric Systems Against Malicious Attacks. *IEEE Trans. Power Syst.* **2017**, *32*, 2953–2965. [CrossRef]
2. Amirioun, M.H.; Aminifar, F.; Shahidehpour, M. Resilience-Promoting Proactive Scheduling Against Hurricanes in Multiple Energy Carrier Microgrids. *IEEE Trans. Power Syst.* **2019**, *34*, 2160–2168. [CrossRef]
3. Wang, K.; Xue, Y.; Guo, Q.; Shahidehpour, M.; Zhou, Q.; Wang, B.; Sun, H. A Coordinated Reconfiguration Strategy for Multi-Stage Resilience Enhancement in Integrated Power Distribution and Heating Networks. *IEEE Trans. Smart Grid* **2023**, *14*, 2709–2722. [CrossRef]
4. Nejat, A.; Solitare, L.; Pettitt, E.; Mohsenian-Rad, H. Equitable Community Resilience: The Case of Winter Storm Uri in Texas. *Int. J. Disaster Risk Reduct.* **2022**, *77*, 103070. [CrossRef]
5. Hurricane Ian after-Action Report. Available online: https://ahca.myflorida.com/content/download/23000/file/After%20Action%20Report%20Hurricane%20Ian.pdf (accessed on 1 August 2023).
6. Manshadi, S.D.; Khodayar, M.E. Resilient Operation of Multiple Energy Carrier Microgrids. *IEEE Trans. Smart Grid* **2015**, *6*, 2283–2292. [CrossRef]
7. Zheng, W.; Lu, H.; Zhang, M.; Wu, Q.; Hou, Y.; Zhu, J. Distributed Energy Management of Multi-entity Integrated Electricity and Heat Systems: A Review of Architectures, Optimization Algorithms, and Prospects. *IEEE Trans. Smart Grid* **2024**, *15*, 1544–1561. [CrossRef]
8. Yan, M.; He, Y.; Shahidehpour, M.; Ai, X.; Li, Z.; Wen, J. Coordinated Regional-District Operation of Integrated Energy Systems for Resilience Enhancement in Natural Disasters. *IEEE Trans. Smart Grid* **2019**, *10*, 4881–4892. [CrossRef]
9. Li, G.; Yan, K.; Zhang, R.; Jiang, T.; Li, X.; Chen, H. Resilience-Oriented Distributed Load Restoration Method for Integrated Power Distribution and Natural Gas Systems. *IEEE Trans. Sustain. Energy* **2022**, *13*, 341–352. [CrossRef]
10. Lin, Y.; Chen, B.; Wang, J.; Bie, Z. A Combined Repair Crew Dispatch Problem for Resilient Electric and Natural Gas System Considering Reconfiguration and DG Islanding. *IEEE Trans. Power Syst.* **2019**, *34*, 2755–2767. [CrossRef]
11. Pan, Z.; Guo, Q.; Sun, H. Feasible Region Method Based Integrated Heat and Electricity Dispatch Considering Building Thermal Inertia. *Appl. Energy* **2017**, *192*, 395–407. [CrossRef]

12. Wang, Y.; Xu, Y.; He, J.; Liu, C.-C.; Schneider, K.P.; Hong, M.; Ton, D.T. Coordinating Multiple Sources for Service Restoration to Enhance Resilience of Distribution Systems. *IEEE Trans. Smart Grid* **2019**, *10*, 5781–5793. [CrossRef]
13. Khatibi, M.; Bendtsen, J.D.; Stoustrup, J.; Molbak, T. Exploiting Power-to-Heat Assets in District Heating Networks to Regulate Electric Power Network. *IEEE Trans. Smart Grid* **2021**, *12*, 2048–2059. [CrossRef]
14. Mao, D.; Wang, P.; Wang, W.; Ni, L. Reliability Segment Design in Single-Source District Heating Networks Based on Valve Network Models. *Sustain. Cities Soc.* **2020**, *63*, 102463. [CrossRef]
15. Akaber, P.; Moussa, B.; Debbabi, M.; Assi, C. Automated Post-Failure Service Restoration in Smart Grid Through Network Reconfiguration in the Presence of Energy Storage Systems. *IEEE Syst. J.* **2019**, *13*, 3358–3367. [CrossRef]
16. Yin, H.; Wang, Z.; Liu, Y.; Qudaih, Y.; Tang, D.; Liu, J.; Liu, T. Operational Reliability Assessment of Distribution Network With Energy Storage Systems. *IEEE Syst. J.* **2023**, *17*, 629–639. [CrossRef]
17. Lei, S.; Chen, C.; Song, Y.; Hou, Y. Radiality constraints for resilient reconfiguration of distribution systems: Formulation and application to microgrid formation. *IEEE Trans. Smart Grid* **2020**, *9*, 3944–3956. [CrossRef]
18. Lei, S.; Wang, J.; Hou, Y. Remote-controlled switch allocation enabling prompt restoration of distribution systems. *IEEE Trans. Power Syst.* **2018**, *5*, 3129–3142. [CrossRef]
19. Arif, A.; Cui, B.; Wang, Z. Switching Device-Cognizant Sequential Distribution System Restoration. *IEEE Trans. Power Syst.* **2022**, *1*, 317–329. [CrossRef]
20. Li, Z.; Su, S.; Jin, X.; Xia, M.; Chen, Q.; Yamashita, K. Stochastic and Distributed Optimal Energy Management of Active Distribution Networks Within Integrated Office Buildings. *CSEE J. Power Energy Syst.* **2024**, *10*, 504–517.
21. Li, Z.; Xu, Y.; Wang, P.; Xiao, G. Coordinated preparation and recovery of a post-disaster Multi-energy distribution system considering thermal inertia and diverse uncertainties. *Appl. Energy* **2023**, *336*, 120736. [CrossRef]
22. Ding, B.; Li, Z.; Li, Z.; Xue, Y.; Chang, X.; Su, J.; Jin, X.; Sun, H. A CCP-based distributed cooperative operation strategy for multi-agent energy systems integrated with wind, solar, and buildings. *Appl. Energy* **2024**, *365*, 123275. [CrossRef]
23. Ding, Y.; Shao, C.; Hu, B.; Bao, M.; Niu, T.; Xie, K.; Singh, C. Operational Reliability Assessment of Integrated Heat and Electricity Systems Considering the Load Uncertainties. *IEEE Trans. Smart Grid* **2021**, *12*, 3928–3939. [CrossRef]
24. Sarantakos, I.; Zografou-Barredo, N.-M.; Huo, D.; Greenwood, D. A Reliability-Based Method to Quantify the Capacity Value of Soft Open Points in Distribution Networks. *IEEE Trans. Power Syst.* **2021**, *36*, 5032–5043. [CrossRef]
25. Wang, K.; Xue, Y.; Zhou, Y.; Zening, L.; Xinyue, C.; Sun, H. Distributed coordinated reconfiguration with soft open points for resilience-oriented restoration in integrated electric and heating systems. *Appl. Energy* **2024**, *365*, 123207. [CrossRef]

electronics

MDPI

Article

Economic Dispatch of Integrated Electricity–Heat–Hydrogen System Considering Hydrogen Production by Water Electrolysis

Jinhao Wang [1], Zhaoguang Pan [2,*], Huaichang Ge [2], Haotian Zhao [2], Tian Xia [2] and Bin Wang [2]

[1] Electric Power Research Institute of State Grid Shanxi Electric Power Company, Taiyuan 030024, China; 7wjh@163.com

[2] Department of Electrical Engineering, Tsinghua University, Beijing 100084, China; 13261454982@163.com (H.G.); zhaohaotian@tsinghua.edu.cn (H.Z.); summersummer@tsinghua.edu.cn (T.X.); wb1984@tsinghua.edu.cn (B.W.)

* Correspondence: panzg09@163.com

Abstract: Water electrolysis is a clean, non-polluting way of producing hydrogen that has seen rapid development in recent years. It offers the possibility of resolving the issue of excessive carbon emissions in conventional hydrogen production methods. In addition, waste heat recovery in hydrogen fuel cells can significantly increase the efficiency of energy use. Thus, to combine the electric power system, the hydrogen energy system, and the district heating system, this research suggests a novel optimal multi-energy complementary electricity–hydrogen–heat model. Rooftop photovoltaics, energy storage batteries, electric boilers, and hydrogen energy systems made up of hydrogen generation, hydrogen storage, and hydrogen fuel cells are all included in the suggested model. Furthermore, the electricity–hydrogen–heat system can be connected successfully using waste heat recovery in hydrogen fuel cells to create a coordinated supply of heat and power. In this work, the waste heat of hydrogen fuel cells is taken into account to increase the efficiency of energy use. To show the effectiveness of the suggested optimal multi-energy complementary model, many case studies have been conducted.

Keywords: integrated energy system; coal to hydrogen; hydrogen production by water electrolysis; low carbon operation

check for updates

Citation: Wang, J.; Pan, Z.; Ge, H.; Zhao, H.; Xia, T.; Wang, B. Economic Dispatch of Integrated Electricity–Heat–Hydrogen System Considering Hydrogen Production by Water Electrolysis. *Electronics* **2023**, *12*, 4166. https://doi.org/10.3390/electronics12194166

Academic Editors: Carlos Andrés García-Vázquez, Jiajia Yang, Zhengmao Li, Hanqing Yang and Zening Li

Received: 16 August 2023
Revised: 18 September 2023
Accepted: 18 September 2023
Published: 7 October 2023

1. Introduction

Due to the significant environmental problems brought on by the widespread use of fossil fuels, a number of sectors have started to explore low-carbon operating solutions to lower carbon emissions in recent years. The chemical industry, a high-carbon industry, has long been the focus of attention for its carbon emission issue, because the traditional method of producing hydrogen stems from the heavy usage of fossil fuels (such as coal) in the process [1,2].

Because of its pollution-free qualities, hydrogen production by water electrolysis has become more and more popular and is gradually becoming an indispensable technology for green hydrogen preparation [3]. The complementary operation of electricity and hydrogen can be realized by closely coupling the electric power system with the hydrogen energy system to build an electricity–hydrogen integrated energy system. It is possible to realize the complementary electricity–hydrogen operation to reduce energy loss and improve energy utilization [4,5]. A composite fuel cell membrane based on a highly proton-conductive, thermally crosslinkable phenylene sulfonic acid copolymer was prepared by pore filling to improve the efficiency of fuel cells in [3]. Zhang et al. proposed a multi-renewable-to-hydrogen production method for renewable-dominated hydrogen fueling stations [6]. A three-phase hybrid rectifier control method was proposed for high-power electrolytic hydrogen production application in [7]. Meng et al. proposed a hybrid rectifier coordinative control method for realizing multiscale frequency regulation [8].

In addition, Sharma et al. focused on photoactive, stable, and cost-effective materials, presenting the photoelectrochemical characterization of a phosphorus–nitrogen-doped carbon material to study carbon-based materials [9]. In [10,11], an analytical approach based on the least square error method is proposed for estimating the model parameters of the proton exchange membrane under various operating conditions.

To lessen the carbon dioxide emissions of the electricity grid, a new clean energy source called the hydrogen energy system can be deployed. Because it combines the benefits of electricity and hydrogen energy, the integrated electricity–hydrogen energy system is a novel idea. In addition to realizing a clean and efficient distributed energy supply and creating an energy internet with electricity at its core, it is a crucial step toward realizing the stable operation of a significant share of modern energy power systems. Pan et al. proposed a planning model for an electricity–hydrogen energy system considering hydrogen production and storage technologies, thereby solving the problem of randomness and strong seasonal fluctuation of new energy power output [12]. An optimal integrated electric power and hydrogen system strategy that utilizes hydrogen tube trailers for transportation was proposed [13,14]. A hydrogen energy storage system comprising alkaline electrolysis, a proton-exchange membrane fuel cell stack, and a high-pressure hydrogen storage tank with a compressor was proposed [15–17]. A multi-stage co-planning model for the power distribution system and hydrogen energy system was developed. Flexible conversion between green power and green hydrogen was realized by the integration of hydrogen energy production, storage, power output, and hydrogenation [16–18]. Zhang et al. developed a planning model of hydrogen refueling stations to maximize long-term profitability [19]. Considering hydrogen trading and long-term hydrogen storage, Weiming proposed a new method for the capacity allocation of hydrogen energy collection systems in industrial parks [20]. Wang Dong considered the demand response (DR) of HFCV and studied the operation of the integrated hydrogen system (IPHS) [21]. A coordinated planning model for power system output and transmission (GT) and HSC with transportable seasonal hydrogen storage was proposed in [22]. In order to realize scientific planning of the primary energy supply structure under the concept of carbon neutrality, Pan Xia established a coupled energy and power optimization planning model of electricity–heat–hydrogen–carbon [23]. However, the heat energy generated by hydrogen fuel cells during operation was not considered in the above study. This heat energy can be converted into electricity through a waste heat recovery system, which can improve the overall energy conversion efficiency of the hydrogen fuel cell stack.

This paper aims to develop an economic dispatch strategy of an integrated electric–heat–hydrogen system considering hydrogen production by water electrolysis to achieve a heat-power coordinated supply that can effectively couple the electricity–hydrogen–heat system.

(i) The proposed model optimizes the allocation of electricity output between rooftop photovoltaics and energy storage batteries to achieve a balance between supply and demand.
(ii) The model determines the optimal operation strategy for the integrated electric–heat–hydrogen system to minimize the total cost, taking into account the hydrogen production by water electrolysis.
(iii) The effectiveness of the proposed economic dispatch strategy is verified through simulation studies, demonstrating its potential in achieving an efficient and coordinated supply of electricity and heat in an integrated energy system.

In Section 2, an overview of the proposed electricity–hydrogen–heat multi-energy complementary optimal model is provided. In Section 3, a multi-energy complementary optimal model is presented, which integrated the electric power system and the hydrogen energy system considering the waste heat recovery of hydrogen fuel cells. The results of testing on the simulation system are presented in Section 4. Finally, Section 5 concludes the paper and suggests future work.

2. Electric–Hydrogen–Heat Integrated Energy Systems

To achieve low-carbon operation in industry parks (shown in Figure 1), a low-carbon operation framework of the integrated electric–heat–hydrogen system considering hydrogen production by water electrolysis is proposed. The electricity–hydrogen–heat multi-energy system consists of three closely linked subsystems: the electric power system, the hydrogen energy system and the district heating system. Three subsystems are connected through coupling components (e.g., hydrogen production devices, electric boilers and hydrogen fuel cells). There are two sources of electricity energy in the electric energy system: a power grid and rooftop photovoltaics. The generated electricity is used for electric load, electric storage device and electrolysis of water to produce hydrogen. There are two sources of hydrogen energy in the hydrogen energy system: electrolytic water hydrogen production devices and coal-based hydrogen production devices. The generated hydrogen is used for hydrogen fuel cells and hydrogen storage devices. There are three sources of heating energy in the district heating system: electric boilers, combined heat and power (CHP) units and the waste heat of hydrogen fuel cells. The generated hydrogen is used for heating the load and heating the storage device.

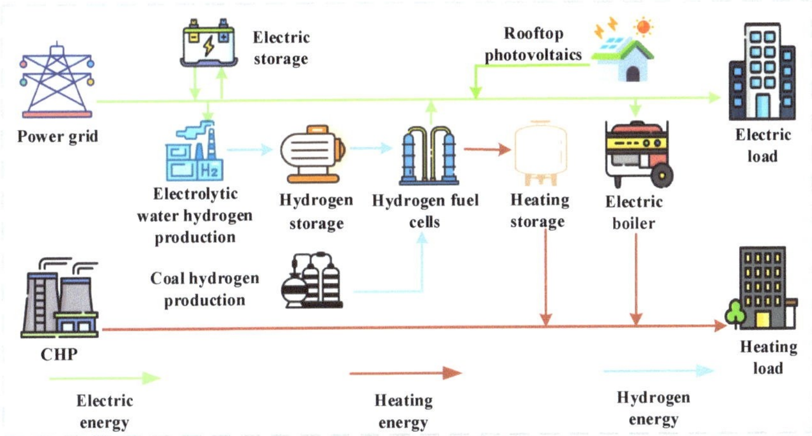

Figure 1. Electricity–hydrogen–heat multi-energy system structure.

In this paper, the key links and coupling equipment in the electric–hydrogen–heat integrated energy systems are modeled, including the hydrogen source, hydrogen load, energy storage equipment and heat source. The electrolytic water hydrogen production equipment and hydrogen fuel cell are connected to the electric power system and hydrogen energy system. The electric energy generated by the power grid and rooftop photovoltaics is used in producing hydrogen energy in the electrolytic water hydrogen production equipment in the electric–hydrogen–heat integrated energy systems. In the electric power system, the electrolytic water hydrogen production equipment can be regarded as an electricity load. In the hydrogen energy system, it can be regarded as a hydrogen source.

In addition, the waste heat recovery of the hydrogen fuel cell is also connected to the district heating system, the electric power system and the hydrogen energy system. The hydrogen energy generated by electrolytic water hydrogen production devices and coal-based hydrogen production devices is used in producing electric energy. Meanwhile, the waste heat of the hydrogen fuel cell is recovered in the hydrogen fuel cell in the electric–hydrogen–heat integrated energy systems. In the electric power system, the electrolytic water hydrogen production equipment can be regarded as an electricity load. In the district heating system, the waste heat recovery of the hydrogen fuel cell can be regarded as a heating source. In the electric power system, it can be regarded as an electric source. In the hydrogen energy system, it can be regarded as a hydrogen load. Furthermore, the electric boiler connects the electric power system and the district heating system. The electric energy generated by the electric source is used in producing heating energy in the electric boiler. In the electric power system, the electric boilers can be regarded as an electric load. In the district heating energy system, it can be regarded as a heating source.

2.1. *Hydrogen Source*

2.1.1. Coal-to-Hydrogen Model

Coal reacts with water vapor at high temperature and high pressure to produce hydrogen. In this process, hydrogen production has a linear relationship

$$m_t^{CTG} = \eta_C m_t^C, t \in T, \tag{1}$$

$$\sum_{t=1}^{24} m_t^C = M^C, t \in T, \tag{2}$$

The carbon emissions generated in the process of coal hydrogen production can be expressed as follows:

$$m_t^{CO_2} = \eta_C m_t^{CTG}, t \in T, \tag{3}$$

$$\sum_{t=1}^{24} m_t^{CO_2} = M^{CO_2}, t \in T, \tag{4}$$

2.1.2. Hydrogen Production from Water Electrolysis Model

Hydrogen production from water electrolysis converts the alternating current of the power grid into direct current through rectification, and it decomposes the water to produce hydrogen through the electrolytic cell. The process can be expressed as follows:

$$P_t^{DC} = \eta_1 P_t^{AC}, m_t^{PTG} = \frac{\eta_2 P_t^{DC} \rho_0}{H_{LHV}}, t \in T, \tag{5}$$

2.2. *Hydrogen Fuel Cell Model*

The waste heat recovery of hydrogen fuel cells can greatly improve energy efficiency and achieve higher output. According to the rule of conservation of energy, its power output and hydrogen consumption conform to the following relationship:

$$P_t^{HFC} = \frac{\eta^{FC} m_t^{HFC} H_{HHV}}{\rho_0}, t \in T \tag{6}$$

In addition, the power output and heat production of hydrogen fuel cell conform to the following linear relationship:

$$h_t^{HFC} = a_h P_t^{HFC} + b_h, t \in T \tag{7}$$

2.3. *Energy Storage Model*

Energy storage equipment in the park includes energy storage battery, hydrogen storage tank and heat storage tank. The energy storage equipment model focuses on the storage change process of energy storage equipment, so the energy storage model of the park can be expressed as follows:

$$E_t^{stor} = E_{t-1}^{stor} + \Delta t(P_{t-1}^{cha} - P_{t-1}^{dis}), t \in T, \tag{8}$$

$$M_t^{stor} = M_{t-1}^{stor} + \Delta t(m_{t-1}^{cha} - m_{t-1}^{dis}), t \in T, \tag{9}$$

$$H_t^{stor} = H_{t-1}^{stor} + \Delta t(h_{t-1}^{cha} - h_{t-1}^{dis}), t \in T, \tag{10}$$

2.4. Heat Source

The heat sources include hydrogen fuel cells and electric boilers, of which the electric boiler is a device that directly converts electric energy into heat energy, and there is a linear conversion relationship between electric heat, so its model can be expressed as follows:

$$h_t^G = \eta^G P_t^G, t \in T, \tag{11}$$

3. Low-Carbon Economic Scheduling Model

3.1. Objective Function

In order to minimize park operating costs and reduce carbon emissions, this paper establishes the following objective functions:

$$\min[\sum_{t=1}^{24} c_t^E P_t^t + c^m M^C + c^c M^{CO_2}] \tag{12}$$

3.2. Operation Constraints

3.2.1. PDS Operation Constraints

A model contains power balance constraints, load-shedding constraints, transmission capacity constraints, voltage drop constraints, and unit output constraints.

1. Power Balance Constraints

$$p_{j,t} = \sum_{s \in \delta(j)} p_{js,t} - \sum_{i \in \pi(j)} (p_{ij,t} - r_{ij} l_{ij,t}), \forall j \in k^{bus}, \forall t \in T, \tag{13}$$

$$q_{j,t} = \sum_{s \in \delta(j)} q_{js,t} - \sum_{i \in \pi(j)} (q_{ij,t} - x_{ij} l_{ij,t}), \forall j \in k^{bus}, \forall t \in T, \tag{14}$$

$$p_{j,t} = p_{j,t}^t + p_{j,t}^{PV} + p_{j,t}^{dis} + p_{j,t}^{HFC} - \left(p_{j,t}^D + p_{j,t}^{AC} + p_{j,t}^G + p_{j,t}^{cha}\right), \forall j \in k^{bus}, \forall t \in T, \tag{15}$$

$$q_{j,t} = q_{j,t}^t + q_{j,t}^{PV} + q_{j,t}^{dis} + q_{j,t}^{HFC} - \left(q_{j,t}^D + q_{j,t}^{AC} + q_{j,t}^G + q_{j,t}^{cha}\right), \forall j \in k^{bus}, \forall t \in T, \tag{16}$$

$$\left\| 2p_{ij,t}\ 2q_{ij,t}\ l_{ij,t} - u_{i,t} \right\|_2 \le l_{ij,t} + u_{i,t}, \forall j \in k^{bus}, \forall t \in T, \tag{17}$$

2. Transmission Capacity Constraints

$$-\overline{S}_{ij} \le p_{ij,t} \le \overline{S}_{ij}, \forall (i,j) \in k^{line}, \forall t \in T, \tag{18}$$

$$-\overline{S}_{ij} \le q_{ij,t} \le \overline{S}_{ij}, \forall (i,j) \in k^{line}, \forall t \in T, \tag{19}$$

3. Voltage Drop Constraints

As shown in Figure 2, the bus voltage will drop along the closed transmission line affected by resistance and reactance, and the bus voltage will not affected by any disconnected bus.

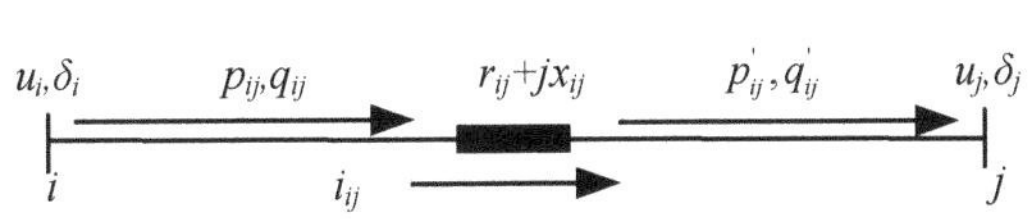

Figure 2. Voltage drop along a transmission line.

The voltage drop constraints will be described as follows:

$$u_{i,c,t} - u_{j,c,t} - 2(r_{ij} p_{ij,c,t} + x_{ij} q_{ij,c,t}) + \left(r_{ij}^2 + x_{ij}^2\right) l_{ij,c,t} = 0, \forall (i,j) \in k^{line}, \forall t \in T, \tag{20}$$

$$u_{\underline{j}} \leq u_{j,c,t} \leq \overline{u}_j, \forall j \in k^{bus}, \forall t \in T, \tag{21}$$

4. Unit Output Constraints

$$\begin{aligned} E^{stor}_{\min} &\leq E^{stor}_t \leq E^{stor}_{\max}, \\ 0 &\leq p^{cha}_{j,c,t}, p^{dis}_{j,c,t} \leq p^{cd}_{\max}, \end{aligned} \tag{22}$$

$$0 \leq P^G_t \leq P^G_{\max}, \tag{23}$$

3.2.2. DHS Operation Constraints

The available heat quantity in the energy flow model is introduced as an auxiliary variable, i.e., $h_{ij} = cm_{ij}\left(\tau^S_{ij} - \tau^R_{ij}\right)$, and the energy flow model is applied to the service restoration model. It contains heat station constraints, heat transmission constraints, energy balance constraints, unit output constraints and load-shedding constraints.

5. Heat Station Constraints

The relationship between the power and heat output of CHP units is expressed as

$$\begin{aligned} &\sum_{j \in k^{HFC}_k} h^{HFC}_{j,t} + \sum_{j \in k^{EB}_k} h^{EB}_{j,t} + \sum_{j \in k^{CHP}_k} h^{dis}_{j,t} = h^{HS}_{k,t}, \forall k \in k^{HS}, \forall t \in T, \\ &h^{HFC}_{j,t} = h^{cha}_{j,t}, \forall k \in k^{HS}, \forall t \in T, \end{aligned} \tag{24}$$

6. Heat Transmission Constraints

Considering the transmission capacity of heating pipelines, the heat energy transmitted though heating pipelines should be limited to a certain range.

$$h^{P,out}_{ij,t} = h^{P,in}_{ij,t} - h^{loss}_{ij,t}, \forall (i,j) \in k^{pipe}, \forall t \in T, \forall c \in C, \tag{25}$$

$$-\overline{h}^P_{ij} \leq h^{P,in}_{ij,c,t} \leq \overline{h}^P_{ij}, \forall (i,j) \in k^{pipe}, \forall t \in T, \forall c \in C, \tag{26}$$

$$-\overline{h}^P_{ij} \leq h^{P,out}_{ij,c,t} \leq \overline{h}^P_{ij}, \forall (i,j) \in k^{pipe}, \forall t \in T, \forall c \in C, \tag{27}$$

7. Energy Balance Constraints

The heat energy generated by the heat sources (CHP units and heating boilers) must be used for the heating load.

$$\sum_{(j,s) \in S^{pipe-}_j} h^{P,out}_{js,c,t} + \sum_{k \in k^{HS}_j} h^{HS}_{k,c,t} = h^L_{j,c} + \sum_{(i,j) \in S^{pipe+}_j} h^{P,in}_{ij,c,t}, \forall j \in k^{nd}, \forall t \in T, \forall c \in C, \tag{28}$$

8. Unit Output Constraints

The heating power output of the CHP units and heating boilers must be limited to a certain range due to the unit standards.

$$\begin{aligned} H^{stor}_{\min} &\leq H^{stor}_t \leq H^{stor}_{\max}, \\ 0 &\leq h^{cha}_t, h^{dis}_t \leq h^{cd}_{\max}, \end{aligned} \tag{29}$$

3.2.3. HS Operation Constraints

1. Hydrogen Source Constraint

The hydrogen output of the electrolytic water hydrogen production devices and coal-based hydrogen production devices must be limited to a certain range.

$$
\begin{aligned}
0 \le m_t^{CTG} \le m_{\max}^{CTG}, \\
0 \le m_t^{PTG} \le m_{\max}^{PTG},
\end{aligned}
\tag{30}
$$

2. Hydrogen Load Constraints

The hydrogen consumption of the hydrogen load and hydrogen fuel cell should be limited to a certain range.

$$
\begin{aligned}
0 \le m_t^{H_2} \le m_{\max}^{H_2}, \\
0 \le m_t^{HFC} \le m_{\max}^{HFC}
\end{aligned}
\tag{31}
$$

3. Hydrogen Storage Constraints

$$
\begin{aligned}
M_{\min}^{stor} \le M_t^{stor} \le M_{\max}^{stor}, \\
0 \le m_t^{cha}, m_t^{dis} \le m_{\max}^{cd}
\end{aligned}
\tag{32}
$$

4. Energy Balance Constraint

$$
\begin{aligned}
& m_t^{PTG} = m_t^{cha}, \\
& m_t^{CTG} + m_t^{dis} = m_t^{HFC} + m_t^{H_2},
\end{aligned}
\tag{33}
$$

4. Case Studies

4.1. Case Description

The proposed electric–hydrogen–heat integrated energy systems model is evaluated using a modified actual industrial area in China, which consists of electric boilers, electrolytic water hydrogen production devices, coal-based hydrogen production devices, hydrogen fuel cells, thermal power units, rooftop photovoltaics, an electric load, a heating load, electric storage, heating storage, and hydrogen storage, with precise parameters supplied in [24]. The tests were conducted using Matlab R2020a on a computer with an i7-1165G7 CPU and 16 GB of memory.

4.2. Cases Analysis

The electricity price used in this article is the time-of-use electricity price [25]. Tariffs vary according to the time of day when electricity is used, with tariffs usually higher during peak periods and lower during low periods, as shown in Table 1.

Table 1. Time-of-use electricity price.

Time Interval	Time	Price ($)
High hours	11:00–15:00, 19:00–22:00	1.08
Low hours	1:00–7:00, 23:00–24:00	0.36
other	8:00–10:00, 16:00–18:00	0.73

The electric power consumption results of hydrogen production from water electrolysis and hydrogen fuel cell power output are shown in Figure 3, and the operation results of energy storage equipment are shown in Figure 4. Using the time-of-use electricity price, the electrolyzer produces hydrogen at 1:00–7:00 and 23:00–24:00, and it is stored in a hydrogen storage to replace coal to produce hydrogen to supply hydrogen load. The energy storage is charged at 1:00–6:00 and 15:00–19:00, and when the electricity price is high, the energy storage is in a discharge state to supply the load. The hydrogen fuel cell generates electricity at 10:00–14:00, 15:00 and 19:00, and it recovers the waste heat to the heat storage to supply the heat load. The operation results of electric boilers are shown in Figure 5. The electric boilers provide heating at 10:00–14:00, 15:00, and 19:00, which provides heat for the heating load. The consumption of electric load and heating load is shown in Figure 6.

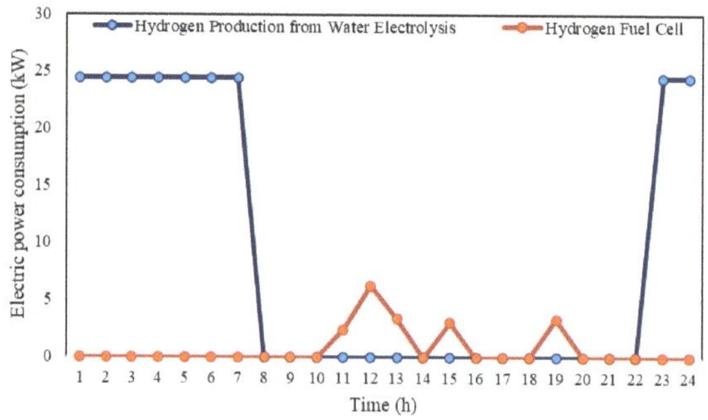

Figure 3. Operation results of electrolyzer and hydrogen fuel cell.

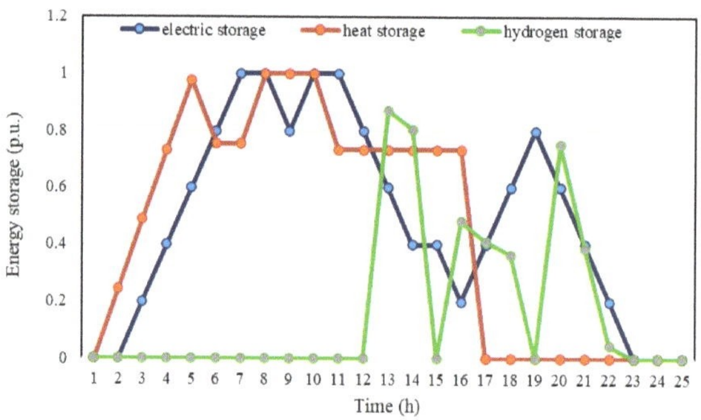

Figure 4. Operation results of energy storage device.

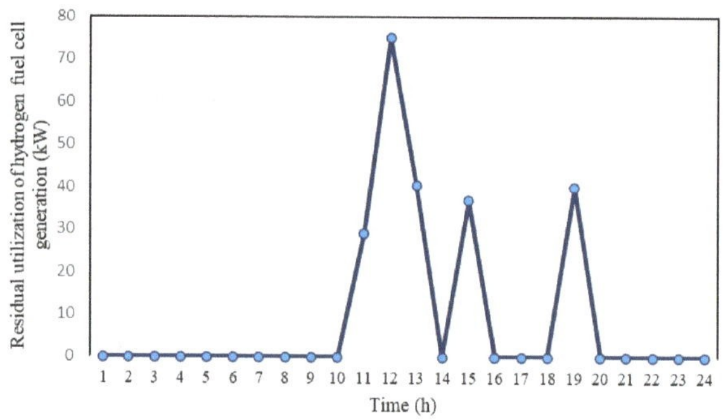

Figure 5. Operation results of electric boiler.

The electricity–hydrogen–heat integrated system realizes the flexible conversion of electricity energy, heat energy and hydrogen energy, which makes full use of the advantages of fast response speed and high energy efficiency of an electric power system and the advantages of suitable energy storage of a heat energy system and hydrogen energy system. Compared to traditional parks that do not introduce photovoltaic, hydrogen, and energy storage batteries, the existence of rooftop photovoltaic, hydrogen energy and energy storage has reduced the daily electricity purchasing cost of the system by $2075, a decrease of 2.4%, reduced coal consumption by 683 kg, a decrease of 1.5%, reduced total cost by $2096, and reduced carbon emissions by 1024 kg. The reason is that the presence of rooftop photovoltaics reduces the power output of thermal power units, thereby reducing the operating costs of thermal power units. Importantly, the carbon emissions can be reduced using the clean energy instead of the fossil energy. The electrolytic water used to produce hydrogen devices reduces the consumption of coal in the process of coal-based hydrogen production and the overall carbon emissions of the system. The role of energy storage is to release energy to ensure user demand when the energy price is high and to store energy when the energy price is low. For example, the electric storage releases electric energy in the period of high electricity price to ensure the consumption of electric load. Charging is performed to store electrical energy when the electricity price is low. Hydrogen storage is also a very large-scale energy storage method which can be used as an electric source for output in buildings, a reliable power source for micro-grids and a backup power source for mobile base stations. However, due to the high operating cost of electrolytic water hydrogen production, the existence of electrolytic water hydrogen production may lead to an increase in the cost of hydrogen energy production. The system operating costs and carbon emissions is shown in Table 2 when the electrolytic cell scale is 600 kW. The effectiveness of the proposed model is verified by the above case.

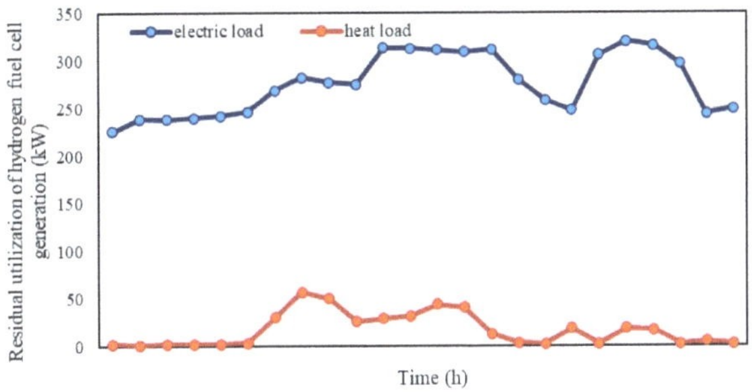

Figure 6. Operation results of electric load and heating load.

Table 2. System operating costs and carbon emissions when electrolytic cell scale is 600 kW.

	Electricity Purchasing Cost ($)	Coal Purchasing Cost ($)	Total Cost ($)	Carbon Emissions (kg)
Traditional model	87,846	40,335	127,597	66,018
Proposed model	85,771	39,739	125,510	64,994

The system operating costs and carbon emissions when the scale of hydrogen production from electrolyzed water in the system is further expanded are shown in Table 3.

Table 3. System operating costs and carbon emissions.

Electrolyzer Scale (kW)	Electricity Purchasing Cost ($)	Coal Purchasing Cost ($)	Total Cost ($)	Carbon Emissions (kg)
600	85,771	39,739	125,510	64,994
800	86,419	39,492	125,911	64,590
1000	87,067	39,245	126,313	64,187
1200	87,715	38,999	126,714	63,784

According to Table 2, it can be observed that carbon emissions will decrease as the electrolytic cell's scale gradually expands. This shows the positive impact of larger electrolytic cells on reducing carbon footprints. There are certain limitations that hinder further expansion of the roof photovoltaic system. One major limitation is the higher electricity purchasing cost compared to the coal purchasing cost. The increasing reliance on electricity purchases would raise the system operating costs, which could have adverse effects on its overall performance and efficiency.

Additionally, the scale of the system also presents a constraint on continual expansion. If the electrolytic cell becomes too large, excessive amounts of electricity would need to be purchased, leading to further increases in system operating costs. These increasing costs could pose challenges to the sustainable and efficient operation of the system. The effectiveness of electric–hydrogen–heat integrated systems has been successfully demonstrated. This suggests that these integrated systems still provide viable solutions for generating clean and renewable energy even if the limitations mentioned above existed.

Moreover, the integration of electric, hydrogen, and heat sources offers diverse benefits, which creates an efficient and interconnected energy framework that allows for the utilization of multiple energy sources simultaneously. By incorporating various energy forms, the system reduces the dependency on conventional fossil fuels and promotes the utilization of cleaner energy. The integration of electric and hydrogen energy contributes to the development of a greener transportation sector, as it enables the use of fuel cells and electric vehicles while minimizing greenhouse gas emissions. Furthermore, the integration of heat sources facilitates the applications in industries that require high-temperature processes, such as steel production or chemical manufacturing.

In conclusion, while the expansion of the electrolytic cell scale leads to a decrease in carbon emissions, some limitations (e.g., high electricity purchasing costs, system size constraints) need to be considered. Nonetheless, the effectiveness and the potential of electric–hydrogen–heat integrated systems in achieving sustainable and clean energy solutions has been proven in this section.

5. Conclusions

In this paper, we propose a novel electricity–hydrogen–heat multi-energy complementary optimal model to integrate the electric power system, the hydrogen energy system and the district heating system. The rooftop photovoltaics, the energy storage batteries devices, the electric boilers, and the hydrogen energy systems consisting of hydrogen production devices, hydrogen storage devices, and hydrogen fuel cells are involved in the proposed model. Moreover, to improve the effectiveness of energy utilization, the waste heat recovery of hydrogen fuel cells is considered in this paper. Numerous case studies are undertaken to demonstrate the effectiveness of a multi-energy complementary optimal model.

For future research, we plan to consider more uncertainties in the electricity–hydrogen–heat multi-energy system. Furthermore, we will explore different ways to improve the effectiveness of energy utilization. In additional, we will also pay attention to coordinated heat and power dispatch (CHPD) with the popularization of CHPD [26–28].

Author Contributions: Conceptualization, B.W.; Methodology, B.W.; Data curation, H.G.; Writing—original draft, Z.P., H.Z. and T.X.; Writing—review & editing, Z.P. and H.Z.; Visualization, H.G.; Supervision, J.W. and H.G.; Project administration, J.W., Z.P. and T.X.; Funding acquisition, J.W. All authors have read and agreed to the published version of the manuscript.

Funding: This research is funded by the Science and Technology Project of Shanxi Electric Power Company No. 52053022000K.

Conflicts of Interest: The authors declare no conflict of interest.

Nomenclature

Sets	
$k_{i,h}^{CHP} / k_{i,e}^{CHP}$	Index of CHP units i in DHS and PDS
k^{pipe} / k^{line}	Set of transmission lines/pipelines
k^{bus} / k^{bus}	Set of buses/nodes
$\pi(j) / \delta(j)$	Set of parents and child buses of bus j
$k^{CHP} / k^{HB} / k^{HS}$	Set of CHP units/heating boilers/and heat stations
$S_j^{pipe-} / S_j^{pipe+}$	Set of pipelines flowing from/to node j
Parameters	
N_s	Number of heat/electric sources
N_{ij}	Number of pipelines/transmission lines ij
r_{ij} / x_{ij}	Binary variable that presents whether there is a fault on line/pipe (i, j)
l_{ij}	Binary variable that presents whether there is a fault on line/pipe (i, j)
$\overline{S}_{ij}$	Transmission capacity of transmission line (i, j) (MW)
$\underline{u}_j / \overline{u}_j$	Minimum/maximum square voltage magnitude of bus j.(V^2)
$\underline{p}_j^{DG} / \overline{p}_j^{DG}$	Minimum/maximum power output of DG j (MW)
$\underline{p}_j^{CHP} / \overline{p}_j^{CHP}$	Minimum/maximum power output of CHP unit j (MW)
$\underline{v}_j / \overline{v}_j$	Minimum/maximum factors of power and heat output of CHP unit j
γ_j	Factors between heat output and fuel consumption of heating boiler j
$\overline{h}_{ij}^{P}$	Maximum transmission capacity of pipelines (i, j)
a_j / b_j	Weight factors for electric and heat load j (kg)
η_C	Conversion efficiency of coal to hydrogen
η_1	Efficiency of the rectification process
η_2	Production efficiency of the electrolytic water hydrogen production
ρ_0	Density of hydrogen in the standard states (kg/m^3)
H_{LHV} / H_{HHV}	Low/high calorific value of hydrogen (J/kg)
η^{FC}	Fuel cell power output efficiency
a_h / b_h	Coefficient of the linear function
η^{G}	Energy conversion efficiency of the electric boiler
c_t^{E}	Real-time electricity price at t ($)
c^{m}	Price of purchasing coal; ($)
c^{c}	Carbon emission penalty coefficient
$E_{\max}^{stor} / E_{\min}^{stor}$	Upper and lower limits of stored electric energy (kWh)
$P_{\max}^{cd}$	Upper limit of charging and discharging power (MW)
$P_{\max}^{G}$	Upper limit of the active power input to the electric boiler (MW)
$H_{\max}^{stor} / H_{\min}^{stor}$	Upper/lower limits of stored heat energy (kWh)
$h_{\max}^{cd}$	Upper limit of heat storage tube (kWh)
$m_{\max}^{CTG} / m_{\max}^{PTG}$	Maximum output of hydrogen production from coal and electrolytic water (kg)
$m_{\max}^{H_2} / m_{\max}^{HFC}$	Upper limits of hydrogen ammonia production and hydrogen fuel cell (kg)
$M_{\max}^{stor} / M_{\min}^{stor}$	Upper and lower limits of stored hydrogen energy (kg)
$m_{\max}^{cd}$	Upper limit of charging and discharging hydrogen flow (kg/s)
Variables	
m_t^{CTG}	Mass flow rate of hydrogen production from coal at t (kg)
m_t^{C}	Mass flow rate of raw coal input at t (kg)
M^{C}	Total coal consumption in a scheduling cycle (kg)
$m_t^{CO_2}$	Carbon emission generated by coal hydrogen production at t (kg)
P_t^{AC} / P_t^{DC}	AC and DC power utilized in the electrolytic water hydrogen production at t (MW)
m_t^{PTG}	Mass flow of hydrogen production at t (kg)
m_t^{HFC}	Mass flow of hydrogen consumed by the hydrogen fuel cell at t (kg)
P_t^{HFC}	Fuel cell power output at t (MW)
h_t^{HFC}	Heat recovered by the hydrogen fuel cell at t (MW)
$E_t^{stor} / E_{t-1}^{stor}$	Stored electric energy at t and $t-1$ (kWh)
$P_{t-1}^{cha} / P_{t-1}^{dis}$	Input and output active power at t and $t-1$ (MW)
$M_t^{stor} / M_{t-1}^{stor}$	Mass of stored hydrogen at t and $t-1$ (kg)
$m_{t-1}^{cha} / m_{t-1}^{dis}$	Input and output hydrogen flow at t and $t-1$ (kg)

$H_t^{stor}/H_{t-1}^{stor}$	Stored heat energy at t and $t-1$ (kWh)
$h_{t-1}^{cha}/h_{t-1}^{dis}$	Input and output of heat at $t-1$ (MW)
P_t^G/h_t^G	Input active power and heat output of the electric boiler at t (MW)
P_t^t	Electricity purchased by the park at t (MW)
$m_{i,c,t}$	Binary variable that presents whether bus i is divided into faulted regions.
$a_{ij,c,t}$	Binary variables that presents the virtual power flow between buses i and j
σ	Binary variable that presents whether SOP is in operation
$m_{i,c,t}$	Binary variable that presents whether bus i is divided into faulted regions.
$p_{j,c,t}/q_{j,c,t}$	Active/Reactive power injection of bus i that is associated with SOP (MW)
$p_{ij,c,t}/q_{ij,c,t}$	Active/Reactive power flow from bus i to bus j (MW)
$p_{j,c,t}^{DG}/q_{j,c,t}^{DG}$	Active/Reactive power output of DG at bus i (MW)
$p_{j,c,t}^{CHP}/q_{j,c,t}^{CHP}$	Active/Reactive power output of CHP unit at bus i (MW)
$p_{j,c,t}^{L}/q_{j,c,t}^{L}$	Electric demand of bus j (MW)
$p_{j,c,t}^{Loss}/q_{j,c,t}^{Loss}$	Load shedding of bus j (MW)
$u_{i,c,t}$	Square voltage of bus j (V)
$h_{j,c,t}^{CHP}$	Heat output of CHP unit j (MW)
$h_{j,c,t}^{HB}/f_{j,c,t}^{HB}$	Heat output and fuel consumption of heating boiler j (MW)
$h_{k,c,t}^{HS}$	Heat output of heat station k (MW)
$h_{ij,c,t}^{P,out}/h_{ij,c,t}^{P,in}/h_{ij,c,t}^{loss}$	Outlet/inlet/loss heat quantity of pipe (i,j) (MW)

References

1. Cai, G.; Kong, L. Techno-economic analysis of wind curtailment/hydrogen production/fuel cell vehicle system with high wind penetration in China. *CSEE J. Power Energy Syst.* **2017**, *3*, 44–52. [CrossRef]
2. Ahmad, J.; Imran, M.; Ali, S.F.; Adnan, M.; Ashraf, S.R.; Hussain, Z.; Shoaib, M. Wind-to-Hydrogen Production Potential for Selected Sites in Pakistan. *IEEE Access* **2021**, *9*, 134874–134898. [CrossRef]
3. Shang, Z.; Hossain, M.M.; Wycisk, R.; Pintauro, P.N. Poly(phenylene sulfonic acid)-expanded polytetrafluoroethylene composite membrane for low relative humidity operation in hydrogen fuel cells. *J. Power Sources* **2022**, *535*, 231375. [CrossRef]
4. Ursua, A.; Gandia, L.M.; Sanchis, P. Hydrogen Production From Water Electrolysis: Current Status and Future Trends. *Proc. IEEE* **2012**, *100*, 410–426. [CrossRef]
5. Iribarren, Á.; Elizondo, D.; Barrios, E.L.; Ibaiondo, H.; Sanchez-Ruiz, A.; Arza, J.; Sanchis, P.; Ursúa, A. Dynamic Modeling of a Pressurized Alkaline Water Electrolyzer: A Multiphysics Approach. *IEEE Trans. Ind. Appl.* **2023**, *59*, 3741–3753. [CrossRef]
6. Zhang, K.; Zhou, B.; Or, S.W.; Li, C.; Chung, C.Y.; Voropai, N. Optimal Coordinated Control of Multi-Renewable-to-Hydrogen Production System for Hydrogen Fueling Stations. *IEEE Trans. Industry Appl.* **2022**, *58*, 2728–2739. [CrossRef]
7. Meng, X.; Chen, M.; He, M.; Wang, X.; Liu, J. A Novel High Power Hybrid Rectifier With Low Cost and High Grid Current Quality for Improved Efficiency of Electrolytic Hydrogen Production. *IEEE Trans. Power Electron.* **2022**, *37*, 3763–3768. [CrossRef]
8. Meng, X.; Jiang, L.; He, M.; Wang, X.; Liu, J. A Novel Multi-Scale Frequency Regulation Method of Hybrid Rectifier and Its Specific Application in Electrolytic Hydrogen Production. *IEEE Trans. Power Electron.* **2023**, *38*, 123–129. [CrossRef]
9. Sharma, R.; Arnoult, K.; Ramasahayam, S.K.; Azam, S.; Hicks, Z.; Shaikh, A.; Viswanathan, T. Photoelectrochemical Hydrogen Production Using Novel Heteroatom-Doped Carbon Under Solar Simulated Radiation. *IEEE Trans. Ind. Appl.* **2016**, *52*, 378–383. [CrossRef]
10. Abomazid, A.M.; El-Taweel, N.A.; Farag, H.E.Z. Novel Analytical Approach for Parameters Identification of PEM Electrolyzer. *IEEE Trans. Ind. Inform.* **2022**, *18*, 5870–5881. [CrossRef]
11. Mazumder, M.; Horenstein, M.N.; Stark, J.W.; Girouard, P.; Sumner, R.; Henderson, B.; Sadder, O.; Hidetaka, I.; Biris, A.S.; Sharma, R. Characterization of Electrodynamic Screen Performance for Dust Removal from Solar Panels and Solar Hydrogen Generators. *IEEE Trans. Ind. Appl.* **2013**, *49*, 1793–1800. [CrossRef]
12. Pan, G.; Gu, W.; Lu, Y.; Qiu, H.; Lu, S.; Yao, S. Optimal Planning for Electricity-Hydrogen Integrated Energy System Considering Power to Hydrogen and Heat and Seasonal Storage. *IEEE Trans. Sustain. Energy* **2020**, *11*, 2662–2676. [CrossRef]
13. Shao, C.; Feng, C.; Shahidehpour, M.; Zhou, Q.; Wang, X.; Wang, X. Optimal Stochastic Operation of Integrated Electric Power and Renewable Energy With Vehicle-Based Hydrogen Energy System. *IEEE Trans. Power Syst.* **2021**, *36*, 4310–4321. [CrossRef]
14. Clegg, S.; Mancarella, P. Integrated Modeling and Assessment of the Operational Impact of Power-to-Gas (P2G) on Electrical and Gas Transmission Networks. *IEEE Trans. Sustain. Energy* **2015**, *6*, 1234–1244. [CrossRef]
15. Li, J.; Li, G.; Ma, S.; Liang, Z.; Li, Y.; Zeng, W. Modeling and Simulation of Hydrogen Energy Storage System for Power-to-gas and Gas-to-power Systems. *J. Mod. Power Syst. Clean Energy* **2023**, *11*, 885–895. [CrossRef]
16. Gómez-Villarreal, H.; Cañas-Carretón, M.; Zárate-Miñano, R.; Carrión, M. Output Capacity Expansion Considering Hydrogen Power Plants and Energy Storage Systems. *IEEE Access* **2023**, *11*, 15525–15539. [CrossRef]
17. Sun, Q.; Wu, Z.; Gu, W.; Liu, P.; Wang, J.; Lu, Y.; Zheng, S.; Zhao, J. Multi-stage Co-planning Model for Power Distribution System and Hydrogen Energy System Under Uncertainties. *J. Mod. Power Syst. Clean Energy* **2023**, *11*, 80–93. [CrossRef]

18. Gong, X.; Dong, F.; Mohamed, M.A.; Abdalla, O.M.; Ali, Z.M. A Secured Energy Management Architecture for Smart Hybrid Microgrids Considering PEM-Fuel Cell and Electric Vehicles. *IEEE Access* **2020**, *8*, 47807–47823. [CrossRef]
19. Zhang, J.; Li, C.; Chen, G.; Dong, Z. Planning of Hydrogen Refueling Stations in Urban Setting While Considering Hydrogen Redistribution. *IEEE Trans. Ind. Appl.* **2022**, *58*, 2898–2908. [CrossRef]
20. Luo, W.; Wu, J.; Cai, J.; Mao, Y.; Chen, S. Capacity Allocation Optimization Framework for Hydrogen Integrated Energy System Considering Hydrogen Trading and Long-Term Hydrogen Storage. *IEEE Access* **2023**, *11*, 15772–15787.
21. Dong, W.; Shao, C.; Feng, C.; Zhou, Q.; Bie, Z.; Wang, X. Cooperative Operation of Power and Hydrogen Energy Systems with HFCV Demand Response. *IEEE Trans. Ind. Appl.* **2022**, *58*, 2630–2639. [CrossRef]
22. Jiang, H.; Qi, B.; Du, E.; Zhang, N.; Yang, X.; Yang, F.; Wu, Z. Modeling Hydrogen Supply Chain in Renewable Electric Energy System Planning. *IEEE Trans. Industry Appl.* **2022**, *58*, 2780–2791. [CrossRef]
23. Xia, P.; Chen, H.; Yuan, B.; Zhang, J.; Wu, C.; Gong, Y. Study on Optimization Method of Energy and Power Path toward Dual Carbon Goal Considering Electricity-Heat-Hydrogen-Carbon Coupling. In Proceedings of the 2023 5th Asia Energy and Electrical Engineering Sympo-Sium, Chengdu, China, 23–26 March 2023; pp. 708–713.
24. Li, J.; Lin, J.; Song, Y.; Xing, X.; Fu, C. Operation Optimization of Power to Hydrogen and Heat (P2HH) in ADN Coordinated With the District Heating Network. *IEEE Trans. Sustain. Energy* **2019**, *10*, 1672–1683. [CrossRef]
25. "Test Data for Economic Dispatch of Integrated Electric-Heat-Hydrogen System Considering Hydrogen Production by Water Electrolysis in Matlab Format". 2023. Available online: https://www.jianguoyun.com/p/DQS4g6UQ8bv3Cxj46ZkFIAA (accessed on 15 August 2023).
26. Wang, K.; Xue, Y.; Guo, Q.; Shahidehpour, M.; Zhou, Q.; Wang, B.; Sun, H. A Coordinated Reconfiguration Strategy for Multi-Stage Resilience Enhancement in Integrated Power Distribution and Heating Networks. *IEEE Trans. Smart Grid.* **2023**, *14*, 2709–2722. [CrossRef]
27. Zhao, L.; Xue, Y.; Sun, H.; Du, Y.; Chang, X.; Su, J.; Li, Z. Benefit allocation for combined heat and power dispatch considering mutual trust. *Appl. Energy* **2023**, *345*, 121279. [CrossRef]
28. Du, Y.; Xue, Y.; Wu, W.; Shahidehpour, M.; Shen, X.; Wang, B.; Sun, H. Coordinated Planning of Integrated Electric and Heating System Considering the Optimal Reconfiguration of District Heating Network. *IEEE Trans. Power Syst.* **2023**, 3242652. [CrossRef]

electronics

Article

Multi-Type Reserve Collaborative Optimization for Gas-Power System Constrained Unit Commitment to Enhance Operational Flexibility

Jinhao Wang [1], Huaichang Ge [2], Zhaoguang Pan [2,*], Haotian Zhao [2], Bin Wang [2] and Tian Xia [2]

[1] Electric Power Research Institute of State Grid Shanxi Electirc Power Company, Taiyuan 030001, China; 7wjh@163.com
[2] Department of Electrical Engineering, Tsinghua University, Beijing 100084, China; 13261454982@163.com (H.G.); zhaohaotian@tsinghua.edu.cn (H.Z.); wb1984@tsinghua.edu.cn (B.W.); summersummer@tsinghua.edu.cn (T.X.)
* Correspondence: panzg09@163.com

Abstract: With the wide application of the gas-power system, gas-power coupling equipment such as gas turbines are gradually becoming widely used, and the problem of insufficient system reserve capacity needs to be solved. In order to improve the operational flexibility of the gas-power system, this paper combines the source side, the load side, and the energy storage side to propose a multi-type backup system, and constructs the source-load-storage multiple reserve capacity system of the gas-power system. Through the participation of gas turbine, steam turbine, interruptible load, and energy storage battery to provide reserve capacity, it can fully cope with the output fluctuation of the load side and source side and realize the coordinated operation of multiple resources to provide reserve capacity. In addition, the coordination of gas turbines and steam turbines can further improve the operation flexibility of the gas-power system. Through the example analysis, it was found that the proposed method reduced the total operating cost of the system by 10.6%.

Keywords: gas-power system; reserve capacity; gas turbine; coordinated operation

Citation: Wang, J.; Ge, H.; Pan, Z.; Zhao, H.; Wang, B.; Xia, T. Multi-Type Reserve Collaborative Optimization for Gas-Power System Constrained Unit Commitment to Enhance Operational Flexibility. *Electronics* **2023**, *12*, 4029. https://doi.org/10.3390/electronics12194029

Academic Editors: Jiajia Yang, Zhengmao Li, Hanqing Yang and Zening Li

Received: 16 August 2023
Revised: 20 September 2023
Accepted: 22 September 2023
Published: 25 September 2023

1. Introduction

In recent years, renewable energy has developed rapidly. At present, China has become the country with the largest installed capacity of wind power in the world [1,2]. By the end of 2022, the cumulative installed capacity of wind power in China was 395.6 W. However, the current wind power prediction level is limited, and the randomness and volatility of wind power output pose challenges to the traditional deterministic scheduling method. At the same time, the frequent natural disasters in recent years may cause multiple power equipment failures, which will lead to power system load loss in severe cases. For example, in March 2022, a blackout occurred in Taiwan, resulting in power outages for about 549 million users [3,4]. The uncertainty of wind power output will further aggravate the harm of forced outage of equipment to power systems. Therefore, the power system needs more flexible scheduling methods to cope with the risks caused by uncertainty.

A generator set reserve is an effective means to ensure the scientific operation of the power system [5]. The traditional unit reserve strategy is to ensure that the total reserve of the system is greater than a certain limit. However, due to the uncertainty of wind power output and the forced outage of power equipment, the reserve strategy cannot meet the requirements of economy and reliability of power system operation [6]. Researchers have conducted a lot of research on the reserve optimization strategy of power systems. In Ref. [7], a reserve supply curve construction algorithm based on robust energy and reserve scheduling is proposed. In Ref. [8], a distributed robust formula based on the concept of conditional value at risk (CVaR) is proposed to solve the reserve demand for wind power. However, the above research only considers how the generator sets provide reserve capacity, which cannot be used in the case of large fluctuations in wind power output.

The gas-power system has been applied more and more in China [9]. By the end of 2022, the number of gas-power systems in China was about 5000. The emergence of gas turbines has increased the power system and natural gas system. A large amount of hot steam will be generated during its operation, which can provide raw materials for steam turbines [10,11]. As a typical device in the gas-power system, the coordinated operation of gas turbines and steam turbines also brings additional operational flexibility to the system. In Ref. [12], in the form of an aggregation model, how to minimize carbon dioxide emissions while providing power and heat loads for gas turbines and steam turbines was studied. Refs. [13,14] establish a configuration-based gas turbine and steam turbine model, which can approximately meet the minimum start–stop constraints of a single turbine and meet other physical constraints of the turbine. In addition, gas turbines and steam turbines together provide spare capacity, which can provide a new method for the flexible operation of gas-power system. Ref. [15] proposes a steam turbine and gas turbine reserve model to solve the uncertainty of renewable energy and realizes the flexible operation of the system. In Ref. [16], the participation of steam turbines and gas turbines in the reserve is considered.

In the gas-power system, in addition to the reserve capacity provided by the source side equipment, the load side and the energy storage side can also provide the reserve capacity for the system [17]. As a kind of power resource to be excavated, interruptible load can be applied to the integrated energy management and microgrid system of the park. The interruptible load on the load side can provide the spinning reserve capacity for the system by reducing the load. The application of energy storage power supply in renewable energy gas-power system provides a feasible solution for the realization of clean and sustainable energy supply [18]. Through the rational design and configuration of the energy storage power system, the reserve capacity of the gas-power system is provided, and the energy management and efficient utilization of resources is guaranteed [19,20].

Therefore, this paper constructs a power generation-load-storage multi-reserve capacity system based on gas power system to solve the problem of system operation fault caused by only the reserve capacity provided by the generator set. The gas power generation system considers gas turbines, steam turbines, interruptible loads, and energy storage batteries to provide reserve capacity. In addition, the cooperative operation of the gas turbine and steam turbine solves the limitation of the traditional gas-power system which only provides flexible resources by gas turbine. The specific contributions of this paper are as follows:

(1) This paper first constructs a source load storage multiple reserve capacity system in the gas-power system, achieving collaborative optimization of multiple types of reserves.
(2) This paper considers the collaborative operation of gas turbines and steam turbines and achieves flexible operation of the system through the combination of different modes.

The structure of the paper is as follows: Section 2 describes the optimal scheduling model of the gas-power system. Section 3 presents the solution strategy of the model. Section 4 describes the case studies. Section 5 is the conclusion.

2. Optimal Scheduling Model of Gas-Power System

Figure 1 shows the architecture of the gas-power system. Among them, gas turbine (GT), steam turbine (ST), and power-to-gas (P2G) units are coupling equipment of power system and natural gas system. GTs and ST are power sources. In this paper, the combination of two GTs and one ST is adopted. Because the operation of a steam turbine mainly relies on the waste heat and smoke generated by the gas turbine, the power output of the steam turbine depends on the operating status of the gas turbine. GT can convert natural gas into electric energy and generate the steam required for ST power generation. The coordinated operation of GT and ST can further improve the flexibility of the system to provide backup. P2G can convert electrical energy into natural gas. In addition, the system also includes energy storage devices and renewable energy.

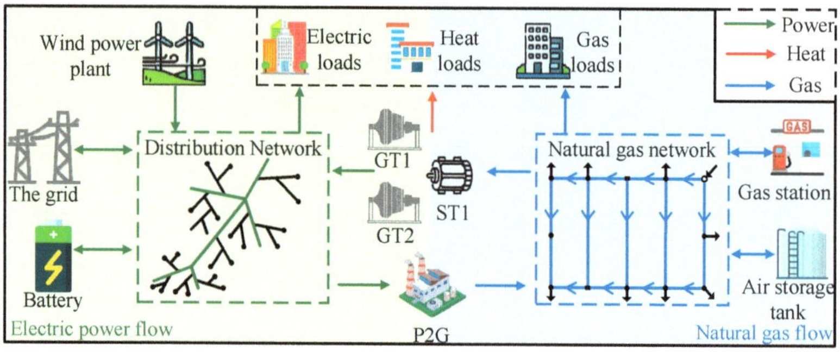

Figure 1. The Structure of gas-power system.

Table 1 shows the operating modes of GTs and ST. The operating states of gas turbine and steam turbine form different operating modes. For example, when a GT runs with 0 STs, the system is in mode 2. When a GT runs with 1 ST, the system is in mode 4. The above operation mode enables the gas turbine and steam turbine to cope with the system reserve capacity demand and operation flexibility demand under the condition of renewable energy output fluctuation and load fluctuation.

Table 1. Different operation modes of GT and ST.

Combination	0 GT + 0 ST	1 GT + 0 ST	2 GT + 0 ST	1 GT + 1 ST	2 GT + 1 ST
Mode	1	2	3	4	5

2.1. Multi-Reserve Capacity System of Gas-Power System

The gas-power system consists of a power generation unit, energy storage device, and interruptible load. Therefore, the gas-power system source-load-storage multiple reserve capacity system can be constructed from the power supply side, the energy storage side, and the load side, as shown in Figure 2. The power supply side mainly relies on GT and ST to provide reserve capacity; the battery in the energy storage device can act as the standby power supply of the system; interruptible load can also provide reserve capacity for the system.

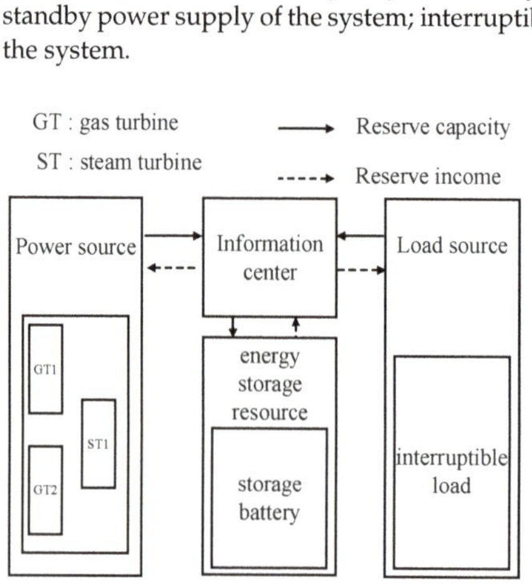

Figure 2. Reserve capacity system of gas-power system.

2.2. Objective Function

The objective function of the model proposed in this paper is to minimize the operating cost of the gas power generation system, including fuel cost, start–stop cost, and reserve capacity benefit.

$$\min \sum_{t\in T}\left(C_{s,t}+C_{f,t}-C_{r,t}^{GT}-C_{r,t}^{L}-C_{r,t}^{ES}\right) \tag{1}$$

where $C_{s,t}$ is the start–stop cost of gas turbines. $C_{f,t}$ is the cost of natural gas used for the gas-power system. $C_{r,t}^{GT}, C_{r,t}^{L}, C_{r,t}^{ES}$ provide reserve income for gas turbines, interruptible load, and energy storage batteries, respectively.

2.2.1. Start–Stop and Operating Costs

$$C_{s,t} = \sum_{i\in\Im^{GT}}\left(c_s^i x_t^i + c_0^i u_t^i\right) \tag{2}$$

where c_s^i, c_0^i are the operating cost and start-up cost of the gas turbines, respectively. x_t^i, u_t^i are the operating binary variables and the starting binary variables of the gas turbines, respectively. $\Im^{GT}$ is the set of gas turbines in the system.

2.2.2. Fuel Costs

$$C^f = \alpha^f \sum_{i\in\Im^{GT}} Q_t^i \tag{3}$$

where α^f is the natural gas price. Q_t^i is the amount of natural gas consumed by the gas turbine.

2.2.3. Reserve Capacity Revenue

$$C_{r,t}^{GT} = \sum_{i\in\Im^{GT}}\left(c_-^i \cdot R_{t,-}^i + c_+^i \cdot R_{t,+}^i\right) \tag{4}$$

$$C_{r,t}^{ES} = \sum_{i\in\Im^{ES}}\left(c_+^i \cdot R_{t,+}^i + c_-^i \cdot R_{t,-}^i\right) \tag{5}$$

$$C_{r,t}^{L} = \sum_{i\in\Im^{L}} c_+^i \cdot R_{t,+}^i \tag{6}$$

where c_-^i, c_+^i represent the benefits of providing down-spin and up-spin reserves. R_-^i, R_+^i represent the down-spinning reserve and up-spinning reserve provided by the equipment. $\Im^{ES}$ is the set of batteries in the system. $\Im^{IL}$ is the set of interruptible loads in the system.

2.3. Operation Constraints of Power System

2.3.1. Constraints of Gas Turbines

Gas turbines are a flexible resource in the gas-power system, which closely connects the power system to the natural gas system. Gas turbines have been increasingly used in electrical coupling systems due to their high electrical efficiency and low carbon emissions. Gas turbines' constraints include power output constraints and start–stop constraints. In addition, gas turbines will generate a large amount of steam during operation, which can drive the steam turbine to generate electricity [21].

$$a\left(P_t^{GT}\right)^2 + bP_t^{GT} + c = H_t Q_t^{GT}, t\in T \tag{7}$$

$$x^{GT}\underline{P}^{GT} \le P_t^{GT} \le x^{GT}\overline{P}^{GT}, t\in T \tag{8}$$

$$-\Delta \underline{P}^{GT} \leq P_t^{GT} - P_{t-1}^{GT} \leq \Delta \overline{P}^{GT}, t \in T \tag{9}$$

$$u_t^{GT} + v_t^{GT} \leq 1, t \in T \tag{10}$$

$$u_{t+1}^{GT} - v_{t+1}^{GT} = x_{t+1}^{GT} - x_t^{GT}, t \in T \tag{11}$$

$$j\left(P_t^{GT}\right)^2 + kP_t^{GT} + l = W_t^{GT}, t \in T \tag{12}$$

$$0 \leq R_{t,-}^{GT} \leq \min\left\{\left(P_t^{GT} - x_t^{GT} \cdot \underline{P}^{GT}\right), x_t^{GT}\left(\tau \cdot \Delta \underline{P}^{GT}\right)\right\} \tag{13}$$

$$0 \leq R_{t,+}^{GT} \leq \min\left\{\left(x_t^{GT} \cdot \overline{P}^{GT} - P_t^{GT}\right), x_t^{GT}\left(\tau \cdot \Delta \overline{P}^{GT}\right)\right\} \tag{14}$$

where a, b, c are the output coefficients of gas turbine, respectively. H_t is the calorific value of natural gas. Q_t^{GT} is the amount of natural gas consumed by the gas turbine. $\underline{P}^{GT}, \overline{P}^{GT}$ are the minimum/maximum output of gas turbine, respectively. $\Delta \overline{P}^{GT}, \Delta \underline{P}^{GT}$ are the uphill and downhill climbing rates of the gas turbine, respectively. v_t^{GT} is the stopping variable of gas turbine. j, k, l are the steam generation coefficient of gas turbine, respectively. W_t^{GT} is the amount of steam produced by the gas turbine. $R_{t,+}^{GT}, R_{t,-}^{GT}$ are the up-spin reserve and down-spin reserve available for gas turbine, respectively.

2.3.2. Constraints of Steam Turbines

The steam turbine (ST) uses the steam generated by the gas turbine to drive the rotor blades to generate electricity. There is also a certain relationship between the steam consumption of steam turbine and the steam production of the gas turbine [22]. The coordinated operation of the gas turbine and steam turbine will further improve the operation flexibility of the gas-power system.

$$P_t^{ST} = f\left(\left(W_t^{ST}\right)^2, W_t^{ST}\right), t \in T \tag{15}$$

$$x^{ST}\underline{P}^{ST} \leq P_t^{ST} \leq x^{ST}\overline{P}^{ST}, t \in T \tag{16}$$

$$-\Delta \underline{P}^{ST} \leq P_t^{ST} - P_{t-1}^{ST} \leq \Delta \overline{P}^{ST}, t \in T \tag{17}$$

$$u_t^{ST} + v_t^{ST} \leq 1, t \in T \tag{18}$$

$$u_{t+1}^{ST} - v_{t+1}^{ST} = x_{t+1}^{ST} - x_t^{ST}, t \in T \tag{19}$$

$$W_t^{ST} + H_t^{HL} \leq W_t^{GT}, t \in T \tag{20}$$

$$0 \leq R_{t,-}^{ST} \leq \min\left\{\left(P_t^{ST} - x_t^{ST} \cdot \underline{P}^{ST}\right), x_t^{ST}\left(\tau \cdot \Delta \underline{P}^{ST}\right)\right\} \tag{21}$$

$$0 \leq R_{t,+}^{ST} \leq \min\left\{\left(x_t^{ST} \cdot \overline{P}^{ST} - P_t^{ST}\right), x_t^{ST}\left(\tau \cdot \Delta \overline{P}^{ST}\right)\right\} \tag{22}$$

where P_t^{ST} is the power output of the steam turbine. (18) and (19) are the relationship between start–stop variables and state variables of steam turbine. (20) indicates that the sum of the steam consumption of the steam turbine and the heat load is less than the steam production of the gas turbine. $R_{t,+}^{ST}, R_{t,-}^{ST}$ are the up-spin reserve and down-spin reserve available for steam turbine, respectively.

2.3.3. Constraints of Batteries

Battery energy storage is a new energy storage technology with high technology maturity at present. It has a fast response speed and the ability to provide fast spinning reserve. The battery plays a role in peak shaving and valley filling in the gas-power system. At the same time, the battery can also provide reserve for the system.

$$V_t = V_{t-1} + \left(\delta^c P_t^c - P_t^{dc}/\delta^{dc}\right)\Delta t \tag{23}$$

$$b^{c/dc}\underline{P}^{c/dc} \le P_t^{c/dc} \le b^{c/dc}\overline{P}^{c/dc}, t \in T \tag{24}$$

$$b_t^c + b_t^{dc} \le 1, t \in T \tag{25}$$

$$\underline{V} \le V_t \le \overline{V}, t \in T \tag{26}$$

$$P_t^{dc} + R_{t,+}^e \le \min\left(\frac{V_t - \underline{V}}{\Delta t}\delta^{dc}, P_0^e\right) \tag{27}$$

$$P_t^c + R_{t,-}^e \le \min\left(\frac{\overline{V} - V_t}{\Delta t \cdot \delta^c}, P_0^e\right) \tag{28}$$

where δ^c, δ^{dc} is the charge/discharge efficiency of the battery. $\underline{P}^{c/dc}, \overline{P}^{c/dc}$ are the battery charge/discharge range. $\overline{V}, \underline{V}$ are the upper/lower limits of battery capacity. $R_{t,+}^e, R_{t,-}^e$ are the up-spin reserve and down-spin reserve available for the battery, respectively. P_0^e is the rated value of battery capacity.

2.3.4. Constraints of Interruptible Load

In the gas-power system, the interruptible load can also provide the spinning reserve for the system to realize the flexible operation of the system.

$$0 \le P_t^{IL} + R_{t,+}^{IL} \le \overline{P}^{IL}, t \in T \tag{29}$$

$$P_t^{IL} \ge 0, R_{t,+}^{IL} \ge 0, t \in T \tag{30}$$

where $\overline{P}^{IL}$ is the maximum power of the interruptible load. $R_{t,+}^{IL}$ is the upper spinning reserve provided by the interruptible load for the system.

2.3.5. Constraints of P2G

P2G is an important part of the gas-power system, which can convert power into natural gas when the electricity price is low, to realize the optimal operation of the system [23].

$$Q_t^g = \delta^g \cdot P_t^g / HHV^g, \forall t \in T \tag{31}$$

$$0 \le P_t^g \le \overline{P}^g, \forall t \in T \tag{32}$$

where δ^g is the power-to-gas efficiency. HHV^g is the high calorific value of natural gas. $\overline{P}^g$ is the maximum power of P2G.

2.3.6. Constraints of Power Balance

The power balance constraint ensures that the system runs in a stable state.

$$\sum_{i \in \Im^{GT} \cap \Im^{ST}} P_t^i + \sum_{w \in \Im^{WD}} P_t^w + \sum \left(P_t^{dc} - P_t^c \right) = \sum_{i \in \Im^{p2g}} P_t^i + P_t^{EL} - P_t^{IL}, \forall t \in T. \tag{33}$$

where $\Im^{WD}$ is the set of wind power plants. $\Im^{p2g}$ is the set of P2Gs. P_t^{EL} is the electric load power.

2.4. *Operation Constraints of Natural Gas System*

The natural gas network model adopts the steady-state Weymouth equation.

$$Q_t^q + \sum_{pq \in X(q)} Q_t^{pq} = \sum_{qk \in Y(q)} Q_t^{qk}, t \in T \tag{34}$$

$$Q_t^q = Q_t^{q,well} + \sum_{g \in \Im_q^{p2g}} Q_t^{q,g} - Q_t^{q,l} - \sum_{i \in \Im_q^{GT}} Q_t^i, t \in T \tag{35}$$

$$Q_t^{pq} = C^{pq} \sqrt{\left| (\varpi^p)^2 - (\varpi^q)^2 \right|}, t \in T \tag{36}$$

where p, q represent the start node and the end node of the natural gas network, respectively. $X(q), Y(q)$ represent the set of starting nodes and ending nodes, respectively. ϖ^p, ϖ^q represent the pressure of the starting node and the ending node, respectively. $Q_t^{q,well}$ is the gas source power output.

3. Solution Methodology

There are quadratic terms of power generation in Equations (7) and (12) in this paper, which can be linearized by cone constraints and introducing auxiliary variables [21]. The Formula (38) is the second-order cone relaxation form of Formula (37).

$$\pi_t = \left(P_t^{GT} \right)^2, t \in T \tag{37}$$

$$\left\| \begin{matrix} \sqrt{2} P_t^{GT} \\ \pi_t \\ 1 \end{matrix} \right\|_2 \le \pi_t + 1, t \in T \tag{38}$$

where π_t is an auxiliary variable introduced by convex quadratic term linearization. The Formula (38) is a linear second-order cone relaxation constraint.

The natural gas system is mainly composed of the natural gas source, natural gas pipeline, and natural gas load. In order to reduce the difficulty of solving the natural gas system model and reduce the complexity of the coupling between the power system and the natural gas system, the nonlinear natural gas pipeline flow equation is linearized by the piecewise linearization method, and then the existing natural gas subsystem model is transformed into a mixed integer linear programming model. The schematic diagram of piecewise linearization is shown in Figure 3.

It should be noted that due to the natural gas in the transmission process by their own materials and external factors, a pressure drop will be produced. In order to maintain the node pressure at a normal level and reduce the probability of gas transmission blockage in the natural gas pipeline during the peak gas load, it is necessary to install compressors along the natural gas pipeline. Because the compressor consumes less energy, in order to simplify the calculation, only the node pressure relationship at both ends of the compressor is retained, and the energy consumed by the compressor is not considered.

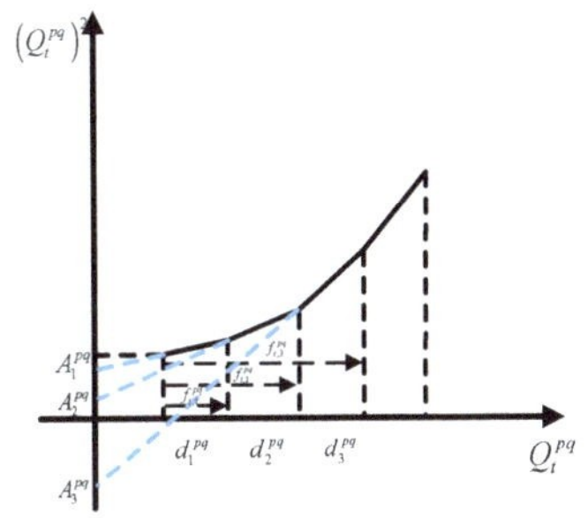

Figure 3. Piecewise linearization diagram of the natural gas network.

$$\left(Q_t^{pq}\right)^2 = (C^{pq})^2 \cdot (\varpi s^p - \varpi s^q), t \in T \tag{39}$$

$$\left(Q_t^{pq}\right)^2 = \sum_{l}^{NL}\left(f_{t,l}^{pq} \cdot \beta_l^{pq} + A_l^{pq} \cdot \delta_l^{pq}\right), t \in T \tag{40}$$

$$0 \le f_{t,l}^{pq} \le \delta_l^{pq} d_l^{pq}, l = 1, t \in T \tag{41}$$

$$\delta_l^{pq} d_{l-1}^{pq} \le f_{t,l}^{pq} \le \delta_l^{pq} d_l^{pq}, l \ge 2, t \in T \tag{42}$$

$$\sum_{l=1}^{NL} \delta_l^{pq} = 1 \tag{43}$$

$$\sum_{l=1}^{NL} f_{t,l}^{pq} = V_t^{pq} \tag{44}$$

$$d_l^{pq} = \frac{Q_{\max}^{pq} l}{m} \tag{45}$$

The linearization process of the natural gas network is shown in Formulas (39)–(45). By dividing the abscissa Q_t^{pq} into m parts, m primary curves are constructed. Formula (39) is the square of the Weymouth equation. Formula (40) is piecewise linearization. The Formulas (41)–(45) are used to judge that the natural gas system works in a certain segmentation interval.

4. Case Studies

All tests were resolved using CPLEX interfaced through MATLAB 2021a [23], and the experimental running computer environment was 11th Gen Intel (R) Core (TM) i7-1165G7 @ 2.80 GHz dual-core processor, 16 GB RAM, and Windows 10 system.

4.1. Test System Description

The example scenario constructed in this paper includes two gas turbines, a steam turbine, a P2G unit, an energy storage battery, and a wind power plant. The purpose of this paper is to study the operation of a gas-power system with multi-type reserve participation. The load, wind power forecasting, and reserve demand in this paper are shown in Figure 4.

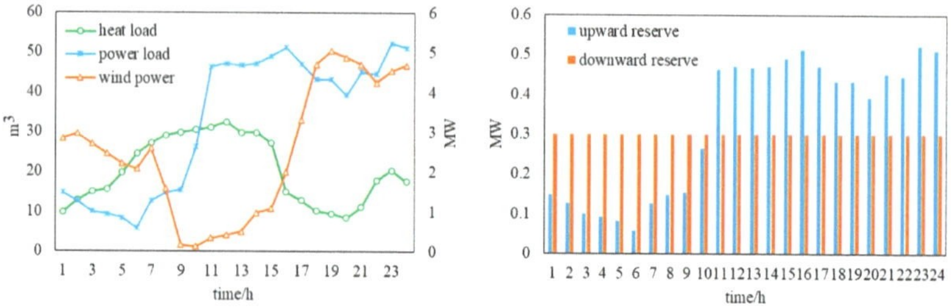

Figure 4. The load, wind power forecasting, and reserve demand.

4.2. Case Setting and Result Analysis

The following three cases are investigated to compare and analyze the results:

Case 1: Gas turbines and steam turbines are involved in providing reserve capacity.

Case 2: Gas turbines, steam turbines, and interruptible loads are involved in providing reserve capacity.

Case 3: Gas turbines, steam turbines, interruptible loads, and energy storage batteries are involved in providing spare capacity.

The scheduling results of the three cases are shown in Table 2.

Table 2. Costs and revenues in case 1 to case 3.

	C_f	C_s	C_r	Total Cost
Case 1	22,718.3	300	618.5	22,399.8
Case 2	22,805.4	300	860.5	21,644.9
Case 3	21,107.3	250	1345.7	20,011.6

It can be seen from the scheduling results that the total cost of the system is reduced when the gas turbine, steam turbine, interruptible load, and energy storage battery participate in the provision of reserve capacity at the same time.

4.2.1. Analysis of Device Power Output

The device power output is shown in Figures 5–7.

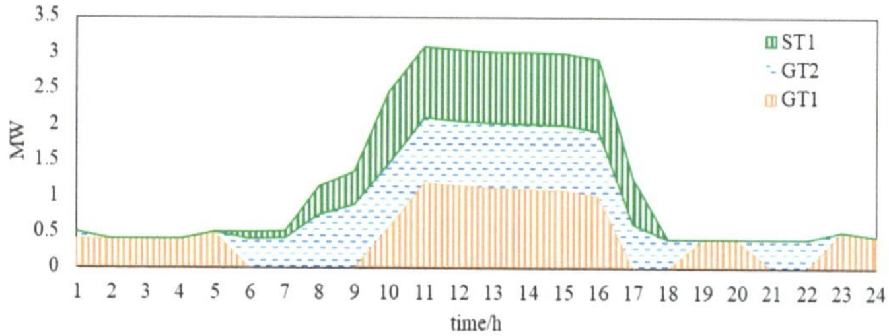

Figure 5. Unit power output in case 1.

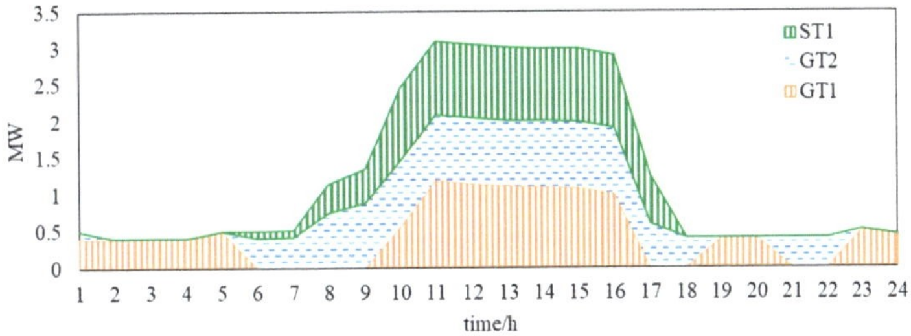

Figure 6. Unit power output in case 2.

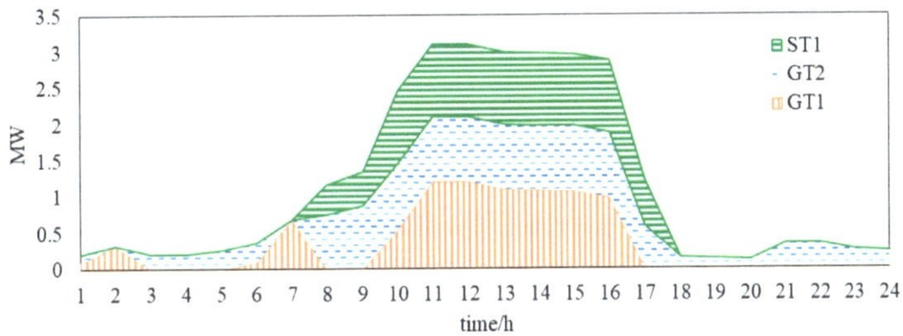

Figure 7. Unit power output in case 3.

From the Figures 5–7, it can be seen that compared with cases 1 and 2, case 3 changes the operation mode of the unit during the period of low load and high wind power. During the period of 3–5, because the power generation cost of GT1 is less than that of GT2, case 1 chooses GT1 to generate power to take into account the reserve demand and heat load demand, and the power generation is higher. In case 3, due to the participation of energy storage batteries in providing backup, the amount of steam generated by GT2 is greater than that of GT1 under the same power generation. Therefore, the small amount of electricity generated by GT2 can meet the heat load demand of the period and take into account the backup demand. Although the power generation cost of GT2 is high, the power generation is small, and the cost is still less than that of case 1. The reasons for the change of unit operation mode in the 19–20 and 23–24 periods are the same as above. In the 6–7 period, the heat load level is high. To meet the load demand and reserve demand, GT2 + ST1 are selected to operate together, and the output of GT2 is at a high level. In case 3, the unit does not need to consider the reserve demand and only needs to select the best operation mode that can meet the electrical load and thermal load at the same time.

In summary, the participation of battery energy storage in reserve makes the total operating hours of the unit less than cases 1 and 2, and the number of start-stops is reduced. At the same time, the total power generation of the unit is reduced, and the fuel cost is reduced, indicating that the participation of battery energy storage in reserve will increase the operational flexibility of the system, making the unit generate electricity in a more economical operation mode.

4.2.2. Analysis of Device Operation Modes

Table 3 shows the operation mode results of GTs and ST in different periods.

Table 3. The device operation mode of case 1 to case 3.

Hour	1	2	3	4	5	6	7	8	9	10	11	12
Mode (Case 1)	4	2	2	2	2	6	6	6	6	7	7	7
Mode (Case 2)	4	2	2	2	2	6	6	6	6	7	7	7
Mode (Case 3)	4	2	3	3	3	4	2	6	6	7	7	7
Hour	**13**	**14**	**15**	**16**	**17**	**18**	**19**	**20**	**21**	**22**	**23**	**24**
Mode (Case 1)	7	7	7	7	6	3	2	2	3	3	2	2
Mode (Case 2)	7	7	7	7	6	3	2	2	3	3	2	2
Mode (Case 3)	7	7	7	7	6	3	3	3	3	3	3	3

It can be seen from the scheduling results that compared with case 1, case 3 changes the operation mode of the unit during low load and high wind power periods.

(1) In the periods of 3–5, because the cost of GT1 power generation is less than that of GT2, case 1 chooses GT1 power generation to take into account the reserve demand and heat load demand, and the power generation is higher. In case 3, due to the participation of energy storage batteries, the amount of steam generated by GT2 is greater than that of GT1 under the same power generation. Therefore, a small amount of electricity generated by GT2 can meet the heat load demand of this period and take into account the reserve demand. Although the power generation cost of GT2 is high, the power generation is small, and the cost is still less than that of case 1. The reasons for the change of unit operation mode in the 19–20 and 23–24 periods are the same as above.

(2) In the 6–7 periods, the heat load level is high. To meet the load demand and reserve demand, case 1 chooses GT2 + ST1 to operate together, and the output of GT2 is at a high level. In the case of case 3, the unit does not need to consider the reserve demand and only needs to select the best operation mode that can meet the electrical load and thermal load at the same time.

In summary, the participation of battery energy storage in reserve makes the total operation hours of the unit less than that of cases 1 and 2, and the number of start-ups and stops is reduced. At the same time, the total power generation of the unit is reduced, and the fuel cost is reduced. It is shown that the participation of battery energy storage in reserve will increase the operation flexibility of GTs and ST, making the generator units operate more economically.

4.2.3. Analysis of Reserve Optimization

The battery capacity changes of the three schemes are shown in Figure 7. The reserve capacity provided by the system is shown in Figure 8.

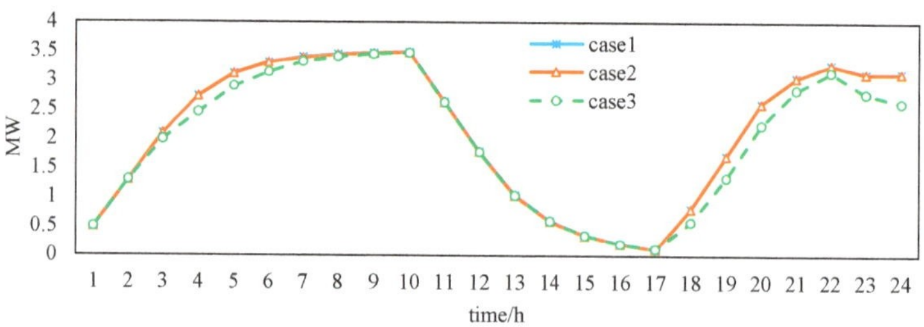

Figure 8. Battery capacity in case 1, 2, and 3.

From the Figures 8 and 9, it can be seen that compared with case 1, case 2 can provide an up-spinning reserve for the system through interruptible load during low load periods, which can increase the output of the unit, make the battery store more electricity during low load period, and reduce the load loss during peak load period.

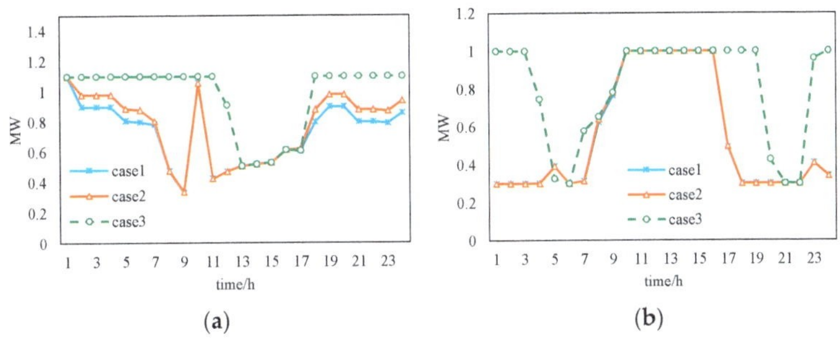

Figure 9. Reserve capacity provided by the system: (**a**) upward spinning reserve, (**b**) downward spinning reserve.

In cases 1 and 2, battery energy storage can only participate in peak shaving operation through low storage and high incidence, and only the generator set increases output charging when the load is low. In case 3, after the energy storage battery participates in the reserve, it will reduce the storage capacity in the low-load 1–6 period to provide the down-spin reserve, so that the unit's output at the 1–6 moment will be reduced. It will also provide the down-spin reserve to reduce the unit's output during the period of high wind power after the 18 period. Therefore, the wind curtailment rate is reduced, and more spinning reserves will be set aside, which increases the reserve income.

5. Conclusions

In this paper, a multi-type reserve capacity system of source-load-storage was constructed based on the gas-power system. Through the example results, when the gas turbine, steam turbine, interruptible load, and energy storage battery participate in the reserve, the operating cost of the system is reduced by 10.6%. In addition, the cooperative operation mode of the gas turbine and steam turbine makes the fuel cost of the system decrease by 7.1%, the start and stop times of the generator set decrease by one time, the overall operation time decrease, and the operation flexibility of the gas-power system increase.

In the future, we will further tap the reserve resources of the gas-power system and improve the operational flexibility of the power system. In addition, we also plan to study the operation of the gas-power system under multiple uncertainties to improve the system's ability to resist risks.

Author Contributions: Conceptualization, B.W. and T.X.; Methodology, Z.P., B.W. and T.X.; Data curation, H.G.; Writing—original draft, Z.P. and H.Z.; Writing—review & editing, H.G., Z.P. and H.Z.; Supervision, J.W. and H.G.; Project administration, J.W.; Funding acquisition, J.W. All authors have read and agreed to the published version of the manuscript.

Funding: This research is funded by the Science and Technology Project of Shanxi Electric Power Company No. 52053022000K.

Conflicts of Interest: The authors declare no conflict of interest.

References

1. Wang, K.; Xue, Y.; Guo, Q.; Shahidehpour, M.; Zhou, Q.; Wang, B.; Sun, H. A Coordinated Reconfiguration Strategy for Multi-Stage Resilience Enhancement in Integrated Power Distribution and Heating Networks. *IEEE Trans. Smart Grid* **2023**, *14*, 2709–2722. [CrossRef]
2. Du, Y.; Xue, Y.; Wu, W.; Shahidehpour, M.; Shen, X.; Wang, B.; Sun, H. Coordinated Planning of Integrated Electric and Heating System Considering the Optimal Reconfiguration of District Heating Network. *IEEE Trans. Power Syst.* 2023; *early access*.
3. Zhang, Q.; Wang, Z.; Ma, S.; Arif, A. Stochastic pre-event preparation for enhancing resilience of distribution systems. *Renew. Sustain. Energy Rev.* **2021**, *152*, 111636. [CrossRef]

4. Zhang, G.; Zhang, F.; Zhang, X.; Meng, K.; Dong, Z.Y. Sequential Disaster Recovery Model for Distribution Systems With Co-Optimization of Maintenance and Restoration Crew Dispatch. *IEEE Trans. Smart Grid* **2020**, *11*, 4700–4713. [CrossRef]
5. Li, Y. New energy utilization in environmental design and realization. *Energy Rep.* **2022**, *8*, 9211–9220.
6. Renewable Capacity Highlights 2020. Available online: https://www.irena.org/statistics (accessed on 2 June 2020).
7. Pan, Z.; Guo, Q.; Sun, H. Interactions of district electricity and heating systems considering time-scale characteristics based on quasi-steady multi-energy flow. *Appl. Energy* **2016**, *167*, 230–243. [CrossRef]
8. Wang, Z.; Bian, Q.; Xin, H.; Gan, D. A Distributionally Robust Co-Ordinated Reserve Scheduling Model Considering CVaR-Based Wind Power Reserve Requirements. *IEEE Trans. Sustain. Energy* **2016**, *7*, 625–636. [CrossRef]
9. Osiadacz, A.J.; Chaczykowski, M. Modeling and Simulation of Gas Distribution Networks in a Multienergy System Environment. *Proc. IEEE* **2020**, *108*, 1580–1595. [CrossRef]
10. Khani, H.; Farag, H.E.Z. Optimal Day-Ahead Scheduling of Power-to-Gas Energy Storage and Gas Load Management in Wholesale Electricity and Gas Markets. *IEEE Trans. Sustain. Energy* **2018**, *9*, 940–951. [CrossRef]
11. Ji, H.; Mi, C.; Yuan, Z.; Liu, Y.; Zhu, H.; Meng, F. Multicomponent Gas Detection Method via Dynamic Temperature Modulation Measurements Based on Semiconductor Gas Sensor. *IEEE Trans. Ind. Electron.* **2023**, *70*, 6395–6404. [CrossRef]
12. Rigo-Mariani, R.; Zhang, C.; Romagnoli, A.; Kraft, M.; Ling, K.V.; Maciejowski, J. A combined cycle gas turbine model for heat and power dispatch subject to grid constraints. *IEEE Trans. Sustain. Energy* **2020**, *11*, 448–456. [CrossRef]
13. Cohen, A.I.; Ostrowski, G. Scheduling units with multiple operating modes in unit commitment. *IEEE Trans. Power Syst.* **1996**, *11*, 497–503. [CrossRef]
14. Colmenar-Santos, A.; Gómez-Camazón, D.; Rosales-Asensio, E.; Blanes-Peiró, J.J. Technological improvements in energetic efficiency and sustainability in existing combined-cycle gas turbine (CCGT) power plants. *Appl. Energy* **2018**, *223*, 30–51. [CrossRef]
15. Tang, Z.; Liu, Y.; Wu, L.; Liu, J.; Gao, H. Reserve Model of Energy Storage in Day-Ahead Joint Energy and Reserve Markets: A Stochastic UC Solution. *IEEE Trans. Smart Grid* **2021**, *12*, 372–382. [CrossRef]
16. Huang, H.; Zhou, M.; Li, G. An Endogenous Approach to Quantifying the Wind Power Reserve. *IEEE Trans. Power Syst.* **2020**, *35*, 2431–2442. [CrossRef]
17. Nosratabadi, S.M.; Hooshmand, R.A.; Gholipour, E. Stochastic profit-based scheduling of industrial virtual power plant using the best demand response strategy. *Appl. Energy* **2016**, *164*, 590–606. [CrossRef]
18. Hong, T.; Zhao, D.; Zhang, Y.; Cui, B.; Tian, Y. Optimal voltage reference for droop-based DERs in distribution systems. *IEEE Trans. Smart Grid* **2020**, *11*, 2357–2366. [CrossRef]
19. Wu, J.; Zhang, B.; Deng, W.; Zhang, K. Application of Cost-CVaR model in determining optimal spinning reserve for wind power penetrated system. *Int. J. Electr. Power Energy Syst.* **2015**, *66*, 110–115. [CrossRef]
20. Dai, W.; Yu, J.; Yang, Z.; Huang, H.; Lin, W.; Li, W. A Static Equivalent Model of Natural Gas Network for Electricity–Gas Co-Optimization. *IEEE Trans. Sustain. Energy* **2020**, *11*, 1473–1482. [CrossRef]
21. Zeng, Z.; Ding, T.; Xu, Y.; Yang, Y.; Dong, Z. Reliability Evaluation for Integrated Power-Gas Systems With Power-to-Gas and Gas Storages. *IEEE Trans. Power Syst.* **2020**, *35*, 571–583. [CrossRef]
22. Gupta, P.; Zan, T.T.T.; Dauwels, J.; Ukil, A. Flow-Based Estimation and Comparative Study of Gas Demand Profile for Residential Units in Singapore. *IEEE Trans. Sustain. Energy* **2019**, *10*, 1120–1128. [CrossRef]
23. He, Y.; Shahidehpour, M.; Li, Z.; Guo, C.; Zhu, B. Robust Constrained Operation of Integrated Electricity-Natural Gas System Considering Distributed Natural Gas Storage. *IEEE Trans. Sustain. Energy* **2018**, *9*, 1061–1071. [CrossRef]

electronics

Article

Collaborative Service Restoration with Network Reconfiguration for Resilience Enhancement in Integrated Electric and Heating Systems

Jinhao Wang [1], Huaichang Ge [2], Yang Yang [1], Zhaoguang Pan [2,*], Yizhao Liu [1] and Haotian Zhao [2]

1 Electric Power Research Institute of State Grid Shanxi Electirc Power Company, Taiyuan 030001, China; 7wjh@163.com (J.W.); miayang_xjtu@163.com (Y.Y.)
2 Department of Electrical Engineering, Tsinghua University, Beijing 100084, China; 13261454982@163.com (H.G.)
* Correspondence: panzg09@163.com

Abstract: Coordinated fault recovery is essential for the resilience enhancement of integrated electric and heating systems (IEHS) following natural catastrophes as the linkage of the power distribution system (PDS) and district heating system becomes tighter. DHS reconfiguration is a viable method for service restoration because it could adjust the energy between energy sources and achieve uninterrupted energy supplies. In this paper, a collaborative service restoration model considering DHS reconfiguration is proposed to achieve better recovery after natural disasters. DHS reconfiguration could guarantee interrupted power supply in non-fault regions by shifting electric loads between power sources and accomplish optimal service restoration by adjusting the power output of combined heat and power units. Numerous case studies are undertaken to demonstrate the performance of coordinated reconfiguration on resilience enhancement and to confirm the efficacy of the proposed paradigm.

Keywords: integrated electric and heating system; collaborative service restoration; DHS reconfiguration; resilience enhancement

Citation: Wang, J.; Ge, H.; Yang, Y.; Pan, Z.; Liu, Y.; Zhao, H. Collaborative Service Restoration with Network Reconfiguration for Resilience Enhancement in Integrated Electric and Heating Systems. *Electronics* **2023**, *12*, 3792. https://doi.org/10.3390/electronics12183792

Academic Editor: Carlos Andrés García-Vázquez

Received: 4 July 2023
Revised: 18 August 2023
Accepted: 19 August 2023
Published: 7 September 2023

1. Introduction

In recent years, frequent natural disasters damaged extensive energy infrastructures and caused massive energy outages [1,2]. In 2012, the superstorm Sandy destroyed the natural gas and power transmission systems in the US [3–5], where more than 4.8 million people suffered natural gas and power outages [6]. In 2020, the ice disaster damaged the power transmission lines. It caused 300 million people to experience energy shortages and forced the unit to shut down at Changchun thermal power plant in Jilin, China.

The resilience of the integrated energy system has garnered a lot of attention as people are becoming more aware of these dangers. The service restoration methods of integrated energy systems have been extensively researched to enhance the integrated energy system resilience after disasters [7,8]. For the purpose of boosting resilience, a technique of service restoration that takes into account the coordinated operation of district and regional integrated energy systems has been proposed [9–12]. A service recovery model was developed for the electric and gas system, which considers subsystem coordination [13]. A repair crew dispatch strategy considering the power distribution system reconfiguration was proposed to enhance the electric and gas system resilience [14].

With the introduction of various coupling elements, such as combined heat and power (CHP) units and heating boilers, the electric and heating systems are now closely related [15,16]. The complicated coupling characteristics between the power distribution system (PDS) and district heating system (DHS) would induce two practical problems: (i) Through coupling components, the defects in PDS/DHS could spread to the other system. (ii) The operation flexibility of CHP units cannot be completely utilized during the fault recovery process when subsystems function independently, according to Khatibi and Liu et al.'s analysis and demonstration of the fault propagation among subsystems [17,18]. Therefore, the joint service restoration approach is required for improving resilience.

DHS reconfiguration is an imperative tool for PDS resilience enhancement [19,20]; nonetheless, it has not been considered in the collaborative recovery process of integrated electric and heating systems (IEHSs) for enhancing the overall system resilience. PDS reconfiguration could achieve uninterruptible power supplies in non-faulted regions and adjust the power generation of CHP units after natural disasters to accomplish better service restoration, which has great potential for resilience enhancement of IEHSs.

This paper presents a collaborative service restoration approach that takes into account the reconfiguration of PDS to enhance the resilience of IEHSs. The proposed method offers several contributions, including the development of a comprehensive framework for service restoration, the consideration of PDS reconfiguration as a means of enhancing IEHS resilience, and the incorporation of collaboration among different stakeholders in the restoration process. Overall, this paper offers a valuable contribution to the field of IEHS resilience and provides a practical approach for enhancing the resilience of these critical systems:

(1) A model for collaborative service restoration is presented, which considers the interaction between the fault isolation and restoration stages. It emphasizes the complex coupling characteristics between PDS and DHS to enhance resilience in park-level IEHSs.
(2) Coordinated reconfiguration is a key focus in the collaborative recovery process of park-level IEHSs. This approach can improve overall system resilience by shifting electric loads between power sources and optimally adjusting power generation of CHP units in PDS to ensure better energy supply during fault recovery progress.

In Section 2, an overview of a park-level IEHS is provided. In Section 3, a comprehensive fault recovery model is presented, which addresses the coordinated reconfiguration during the recovery process. The results of testing on the P33H14 system are presented in Section 4—concludes the paper and discusses future work.

2. A Collaborative Service Restoration Model for Park-Level IEHS

The park-level IEHS consists of two closely linked subsystems: the power distribution system and the district heating system. These two subsystems are connected through coupling components (e.g., CHP units). The CHP units serve as the primary energy source for the DHS and PDS, further strengthening the relationship between the two subsystems.

The process of fault recovery in IEHSs can be broken down into two stages: fault isolation and service restoration. In the fault isolation stage, it has been discovered that IEHS fault isolations cannot be achieved by PDS or DHS operators. Coordinated operation of DHS and PDS can shift partial abnormal nodes/buses to non-faulted regions, reducing initial faulted regions. In the fault restoration stage, reconfigurations of DHN and PDS can be coordinated to recover load shedding in normal regions. It is important to note that flexibility resources are not exploited to enhance park-level IEHS resilience unless they are coordinated together. Coordinated operation of DHS valves and PDS switches is essential for fault isolation and service restoration in IEHSs after disasters to enhance IEHS resilience.

In this section, we present a collaborative service restoration model that takes into account the reconfiguration of the power distribution system. The model comprises fault isolation and restoration models. During the fault isolation stage, the PDS reconfiguration enables uninterrupted power supply in non-faulted regions. In the service restoration stage, the PDS reconfiguration enhances fault recovery by leveraging the operational flexibility of combined heat and power units. This approach offers a promising solution for improving the reliability and resilience of IEHS.

2.1. Topological Constraints

2.1.1. Fault Isolation Model

The fault isolation model is a crucial tool in network management, as it allows for the accurate identification of faulted regions and the description of fault propagation throughout the network. This model is designed to provide network administrators with a comprehensive understanding of the network's behavior in the event of a fault, allowing them to quickly and effectively address any issues that arise [21]. The fault isolation model identifies the faulted regions accurately and describes fault propagation in the network, which is shown as follows:

$$(1-f_{ij,c})(z_{ij,0}-s_{ij,0}) \le z_{ij,c,t} \le (1-f_{ij,c})z_{ij,0}, \forall(i,j) \in k^{pipe} \cup k^{line}, \forall t \in T_i, \forall c \in C, \quad (1)$$

$$m_{i,c,t} - z_{ij,0} + 1 \ge f_{ij,c}(1-s_{ij,0}), \forall(i,j) \in k^{pipe} \cup k^{line}, \forall t \in T_i, \forall c \in C, \quad (2)$$

$$m_{j,c,t} - z_{ij,0} + 1 \ge f_{ij,c}(1-s_{ij,0}), \forall(i,j) \in k^{pipe} \cup k^{line}, \forall t \in T_i, \forall c \in C, \quad (3)$$

$$m_{j,c,t} - z_{ij,c,t} + 1 \ge m_{i,c,t}, \forall(i,j) \in k^{pipe} \cup k^{line}, \forall t \in T_i, \forall c \in C, \quad (4)$$

$$m_{i,c,t} - z_{ij,c,t} + 1 \ge m_{j,c,t}, \forall(i,j) \in k^{pipe} \cup k^{line}, \forall t \in T_i, \forall c \in C, \quad (5)$$

$$m_{g,c,t} = m_{h,c,t}, \forall g \in k_{i,h}^{CHP}, h \in k_{i,e}^{CHP}, \forall t \in T_i, \forall c \in C. \quad (6)$$

where k^{pipe} and k^{line} are the set of lines and pipes, $k_{i,h}^{CHP}$ and $k_{i,e}^{CHP}$ are the set of CHP units i in DHS and PDS, T_i represents the fault isolation period, $z_{ij,0}$ is a binary variable that represents whether the line/pipe (i,j) is closed in the pre-event stage, $s_{ij,0}$ is a binary variable that represents whether the line/pipe (i,j) is equipped with a switch in the pre-event stage, $z_{ij,c,t}$ is a binary variable that represents whether the line/pipe (i,j) is connected in the fault isolation stage during period t, $f_{ij,c}$ is a binary variable that represents whether there is a fault on line/pipe (i,j), and $m_{i,c,t}$ is a binary variable that represents whether bus i is divided into faulted regions.

Constraint (1) implies that the switches/valves in non-faulted regions could be operated for fast fault isolation. Constraints (2) and (3) indicate that when there is a fault occurring on a pipe/line, the nodes/buses of the pipe/line will be involved in the faulted/non-faulted region according to switch/valve configuration. Constraints (4) and (5) indicate that the two nodes/buses of a closed pipe/line will be involved in the same region. Constraint (6) illustrates that if CHP units are faulted in DHS/PDS, they are also faulted in the other subsystem. Overall, the abovementioned constraints provide valuable insights into fault isolation in pipeline systems and can be used to develop effective strategies for identifying and addressing faults quickly and efficiently.

2.1.2. Service Restoration Model

After identifying the fault location, switches and valves will be utilized to restore the lost loads in the unaffected regions. The topological constraints will also be taken into consideration during this process. These constraints are formulated based on the information gathered during the fault isolation stage [22].

$$(1-f_{ij,c})(z_{ij,t-1}-s_{ij,0}) \le z_{ij,c,t} \le (1-f_{ij,c})(z_{ij,t-1}+s_{ij,0}), \forall(i,j) \in k^{pipe} \cup k^{line}, \forall t \in T_r, \forall c \in C, \quad (7)$$

$$a_{ij,c,t} + a_{ji,c,t} = z_{ij,c,t}, \forall(i,j) \in k^{pipe} \cup k^{line}, \forall t \in T_r, \forall c \in C, \quad (8)$$

$$\sum_{i \in \pi(j)} a_{ij,c,t} \le 1, \forall j \in k^{nd}, \forall t \in T_r, \forall c \in C, \quad (9)$$

$$\sum_{s\in\sigma(j)} a_{js,c,t} = 0, \forall j \in k^{nd}, \forall t \in T_r, \forall c \in C, \tag{10}$$

$$z_{ij,c,t} = N_{ij} - N_s, \forall(i,j) \in k^{pipe} \cup k^{line}, \forall t \in T_r, \forall c \in C, \tag{11}$$

$$m_{j,c,t-1} - z_{ij,c,t} + 1 \geq m_{i,c,t-1}, \forall(i,j) \in k^{pipe} \cup k^{line}, \forall t \in T_r, \forall c \in C, \tag{12}$$

$$m_{i,c,t-1} - z_{ij,c,t} + 1 \geq m_{j,c,t-1}, \forall(i,j) \in k^{pipe} \cup k^{line}, \forall t \in T_r, \forall c \in C, \tag{13}$$

where T_r represents the service restoration period, and $a_{ij,c,t}$ and $a_{ji,c,t}$ are binary variables that represent the virtual power flow between buses i and j. When $a_{ij,c,t}$ is one, bus i is the parent of bus j in the spanning tree. N_{ij} is the number of pipes/lines, and N_s is the number of heat/electric sources.

The power distribution network is a critical infrastructure that requires constant monitoring and maintenance to ensure uninterrupted power supply to consumers. In the event of a fault, it is essential to isolate the affected area to prevent further damage and restore power to the non-faulted regions as quickly as possible. Constraint (7) illustrates that the pipes/lines equipped with the switches in non-faulted regions could be switched for network reconfiguration. The topology should be radial, as shown in Constraints (8)–(10). This ensures that there is only one path for power flow, which simplifies fault detection and isolation. In the fault isolation stage, the switches in the non-faulted regions can be used to reconfigure the network and restore power to the affected areas. Constraints (12)–(13) illustrate that the faulted areas will not be reconnected to the non-faulted zones in the fault isolation stage. It is important to ensure that the restoration process is carried out in a safe and efficient manner, while also adhering to the relevant regulations and standards. The use of advanced technologies and tools can help to streamline the restoration process and minimize the impact of the fault on the power system. Additionally, ongoing monitoring and maintenance of the power system can help to prevent future faults and ensure the reliability and stability of the system. Overall, a comprehensive approach to fault management and power system restoration is essential for ensuring the continued operation and success of the power grid.

2.2. *Operation Constraints*

2.2.1. PDS Operation Constraints

A mixed-integer second-order cone programming model is formulated for solving the collaborative service restoration problem in [23–27]. It contains power balance constraints in (14)–(18), transmission capacity constraints in (19) and (20), voltage drop constraints in (21)–(23), unit output constraints in (24)–(27), and load shedding constraints in (28) and (29).

1. Power Balance Constraints

$$p_{j,c,t} = \sum_{s\in\delta(j)} p_{js,c,t} - \sum_{i\in\pi(j)} (p_{ij,c,t} - r_{ij} l_{ij,c,t}), \forall j \in k^{bus}, \forall t \in T, \forall c \in C, \tag{14}$$

$$q_{j,c,t} = \sum_{s\in\delta(j)} q_{js,c,t} - \sum_{i\in\pi(j)} (q_{ij,c,t} - x_{ij} l_{ij,c,t}), \forall j \in k^{bus}, \forall t \in T, \forall c \in C, \tag{15}$$

$$p_{j,c,t} = p^{DG}_{j,c,t} + p^{CHP}_{j,c,t} + \sigma p^{SNOP}_{j,c,t} - \left(p^{L}_{j,c,t} - p^{Loss}_{j,c,t}\right), \forall j \in k^{bus}, \forall t \in T, \forall c \in C, \tag{16}$$

$$q_{j,c,t} = q^{DG}_{j,c,t} + q^{CHP}_{j,c,t} + \sigma q^{SNOP}_{j,c,t} - \left(q^{L}_{j,c,t} - q^{Loss}_{j,c,t}\right), \forall j \in k^{bus}, \forall t \in T, \forall c \in C, \tag{17}$$

$$\left\|2p_{ij,c,t}\ 2q_{ij,c,t}\ l_{ij,c,t} - u_{i,c,t}\right\|_2 \leq l_{ij,c,t} + u_{i,c,t}, \forall j \in k^{bus}, \forall t \in T, \forall c \in C, \tag{18}$$

where k^{bus} is the set of buses; $\pi(j)$ and $\delta(j)$ are the parent and child buses of bus j; $p_{j,c,t}$ and $q_{j,c,t}$ are the power injections of bus j; $p_{ij,c,t}$ and $q_{ij,c,t}$ are the power flow from bus i to bus j; r_{ij} and x_{ij} are resistance and reactance of the line (i,j); l_{ij} is the square current of the line (i,j); $p^{DG}_{j,c,t}$, $p^{CHP}_{j,c,t}$, and $p^{SNOP}_{j,c,t}$ are the power generation of distributed generation (DG), CHP unit, and SNOP; $q^{L}_{j,c,t}$ and $q^{Loss}_{j,c,t}$ are the electric demand and load shedding of bus j; and $u_{i,c,t}$ is the square voltage of bus j.

2. Transmission Capacity Constraints

$$-z_{ij,c,t}\overline{S}_{ij} \leq p_{ij,c,t} \leq z_{ij,c,t}\overline{S}_{ij}, \forall(i,j) \in k^{line}, \forall t \in T, \forall c \in C, \tag{19}$$

$$-z_{ij,c,t}\overline{S}_{ij} \leq q_{ij,c,t} \leq z_{ij,c,t}\overline{S}_{ij}, \forall(i,j) \in k^{line}, \forall t \in T, \forall c \in C, \tag{20}$$

where $\overline{S}_{ij}$ is the transmission capacity of the line (i,j).

3. Voltage Drop Constraints

$$u_{i,c,t} - u_{j,c,t} - 2(r_{ij}p_{ij,c,t} + x_{ij}q_{ij,c,t}) + \left(r_{ij}^2 + x_{ij}^2\right)l_{ij,c,t} \leq (1 - z_{ij,c,t})M, \forall(i,j) \in k^{line}, \forall t \in T, \forall c \in C, \tag{21}$$

$$u_{i,c,t} - u_{j,c,t} - 2(r_{ij}p_{ij,c,t} + x_{ij}q_{ij,c,t}) + \left(r_{ij}^2 + x_{ij}^2\right)l_{ij,c,t} \geq (1 - z_{ij,c,t})M, \forall(i,j) \in k^{line}, \forall t \in T, \forall c \in C, \tag{22}$$

$$\underline{u}_j \leq u_{j,c,t} \leq \overline{u}_j, \forall j \in k^{bus}, \forall t \in T, \forall c \in C, \tag{23}$$

where $\underline{u}_j$ and $\overline{u}_j$ are the minimum and maximum square voltage magnitude of bus j.

4. Unit Output Constraints

$$(1 - m_{j,c,t})\underline{p}_j^{CHP} \leq p_{j,c,t}^{CHP} \leq (1 - m_{j,c,t})\overline{p}_j^{CHP}, \forall j \in k^{CHP}, \forall t \in T, \forall c \in C \tag{24}$$

$$(1 - m_{j,c,t})\underline{q}_j^{CHP} \leq q_{j,c,t}^{CHP} \leq (1 - m_{j,c,t})\overline{q}_j^{CHP}, \forall j \in k^{CHP}, \forall t \in T, \forall c \in C, \tag{25}$$

$$(1 - m_{j,c,t})\underline{p}_j^{DG} \leq p_{j,c,t}^{DG} \leq (1 - m_{j,c,t})\overline{p}_j^{DG}, \forall j \in k^{DG}, \forall t \in T, \forall c \in C, \tag{26}$$

$$(1 - m_{j,c,t})\underline{q}_j^{DG} \leq q_{j,c,t}^{DG} \leq (1 - m_{j,c,t})\overline{q}_j^{DG}, \forall j \in k^{DG}, \forall t \in T, \forall c \in C, \tag{27}$$

where $\underline{p}_j^{CHP}$, $\underline{q}_j^{CHP}$ and $\overline{p}_j^{CHP}$, $\overline{p}_j^{CHP}$ are the limited power generation of CHP unit j; $\underline{p}_j^{DG}$, $\underline{q}_j^{DG}$ and $\overline{p}_j^{DG}$, $\overline{p}_j^{DG}$ are the limited power generation of DG j; Constraints (24)–(27) illustrate that when unit shutdown occurs in the faulted regions, CHP units/DG would not provide the power supply.

5. Load Shedding Constraints

$$m_{j,c,t}p_j^L \leq p_{j,c,t}^{Loss} \leq p_j^L, \forall j \in k^{bus}, \forall t \in T, \forall c \in C, \tag{28}$$

$$m_{j,c,t}q_j^L \leq q_{j,c,t}^{Loss} \leq q_j^L, \forall j \in k^{bus}, \forall t \in T, \forall c \in C. \tag{29}$$

Constraints (28) and (29) illustrate that the electric loads would be fully shed in faulted regions because the unit shutdown and the partial loads in non-faulted regions would be lost for energy balance. It is evident that in the event of a fault in a particular region, the electric loads in that region would be completely shed. This is due to the fact that the unit would shut down, resulting in a loss of partial loads in non-faulted regions, which would disrupt the energy balance of the system.

2.2.2. DHS Operation Constraints

The available heat quantity in the energy flow model is introduced as an auxiliary variable, i.e., $h_{ij} = cm_{ij}\left(\tau_{ij}^{S} - \tau_{ij}^{R}\right)$, and the energy flow model is applied to the service restoration model [28–31]. It contains heat station constraints in (30)–(32), heat transmission constraints in (33)–(35), energy balance constraints in (36), unit output constraints in (37), and load shedding constraints in (38).

1. Heat Station Constraints

CHP units are the main heating sources in industrial parks in China and commonly operate in the mode of determining electricity by heat [20–24]. Thus, the relationship between power and heat generation of CHP units is expressed as

$$\underline{\nu}_j h_{j,c,t}^{CHP} \le p_{j,c,t}^{CHP} \le \overline{\nu}_j h_{j,c,t}^{CHP}, \forall j \in k^{CHP}, \forall t \in T, \forall c \in C, \tag{30}$$

$$h_{j,c,t}^{HB} = \gamma_j f_{j,c,t}^{HB}, \forall j \in k^{HB}, \forall t \in T, \forall c \in C, \tag{31}$$

$$\sum_{j \in k_k^{CHP}} h_{j,c,t}^{CHP} + \sum_{j \in k_k^{HB}} h_{j,c,t}^{HB} = h_{k,c,t}^{HS}, \forall k \in k^{HS}, \forall t \in T, \forall c \in C, \tag{32}$$

where k^{CHP}, k^{HB}, and k^{HS} are the set of CHP units, heating boilers, and heat stations; $h_{j,c,t}^{CHP}$ is the heat generation of CHP unit j; $\underline{\nu}_j$ and $\overline{\nu}_j$ are the minimum and maximum coefficient of power and heat generation of CHP unit j; $h_{j,c,t}^{HB}$ and $f_{j,c,t}^{HB}$ are the heat generation and fuel consumption of heating boiler j; γ_j is the coefficient between heat generation and fuel consumption of heating boiler j; and $h_{k,c,t}^{HS}$ is the heat generation of heat station k.

2. Heat Transmission Constraints

$$h_{ij,c,t}^{P,out} = h_{ij,c,t}^{P,in} - h_{ij,c,t}^{loss}, \forall (i,j) \in k^{pipe}, \forall t \in T, \forall c \in C, \tag{33}$$

$$-z_{ij,c,t}\overline{h}_{ij}^{P} \le h_{ij,c,t}^{P,in} \le z_{ij,c,t}\overline{h}_{ij}^{P}, \forall (i,j) \in k^{pipe}, \forall t \in T, \forall c \in C, \tag{34}$$

$$-z_{ij,c,t}\overline{h}_{ij}^{P} \le h_{ij,c,t}^{P,out} \le z_{ij,c,t}\overline{h}_{ij}^{P}, \forall (i,j) \in k^{pipe}, \forall t \in T, \forall c \in C, \tag{35}$$

where $h_{ij,c,t}^{P,out}$, $h_{ij,c,t}^{P,in}$, and $h_{ij,c,t}^{loss}$ are the outlet heat quantity, inlet heat quantity, and lost heat quantity of the pipe (i,j), and $\overline{h}_{ij}^{P}$ is the limited transmission of the pipe (i,j).

3. Energy Balance Constraints

$$\sum_{(j,s) \in S_j^{pipe-}} h_{js,c,t}^{P,out} + \sum_{k \in k_j^{HS}} h_{k,c,t}^{HS} = h_{j,c}^{L} - h_{j,c,t}^{Loss} + \sum_{(i,j) \in S_j^{pipe+}} h_{ij,c,t}^{P,in}, \forall j \in k^{nd}, \forall t \in T, \forall c \in C, \tag{36}$$

where S_j^{pipe-} and S_j^{pipe+} are the set of pipes flowing from/to node j.

4. Unit Output Constraints

$$(1 - m_{j,c,t})\underline{h}_j^{CHP} \le h_{j,t}^{CHP} \le (1 - m_{j,c,t})\overline{h}_j^{CHP}, \forall j \in k^{CHP}, \forall t \in T, \forall c \in C, \tag{37}$$

Constraint (37) illustrates that when unit shutdown occurs in the faulted regions, CHP units would not provide the heat supply.

5. Load Shedding Constraints

$$m_{j,c,t}h_j^{L} \le h_{j,c,t}^{Loss} \le h_j^{L}, \forall j \in k^{nd}, \forall t \in T, \forall c \in C \tag{38}$$

Constraint (38) illustrates that the heat loads would be fully shed in the faulted regions and partial loads in the non-faulted regions would be lost.

2.2.3. Objective and Resilience Metrics

The objective and resilience metrics are proposed in (39) and (40) in order to reduce the loss of electric and heat loads during the fault recovery process and evaluate the park-level IEHS resilience [32–35].

$$\min \sum_{c\in C} p_c \left\{ T_i\left(\sum_{j\in k^{bus}} a_j p_{j,c,t}^{Loss} + \sum_{j\in k^{nd}} b_j h_{j,c,t}^{Loss} \right) + T_r\left(\sum_{j\in k^{bus}} a_j p_{j,c,t}^{Loss} + \sum_{j\in k^{nd}} b_j h_{j,c,t}^{Loss} \right) \right\}, \quad (39)$$

$$R_c = 1 - \frac{T_i\left(\sum\limits_{j\in k^{bus}} a_j p_{j,c,t}^{Loss} + \sum\limits_{j\in k^{nd}} b_j h_{j,c,t}^{Loss} \right) + T_r\left(\sum\limits_{j\in k^{bus}} a_j p_{j,c,t}^{Loss} + \sum\limits_{j\in k^{nd}} b_j h_{j,c,t}^{Loss} \right)}{T\left(\sum\limits_{j\in k^{bus}} a_j p_{j,c,t}^{Loss} + \sum\limits_{j\in k^{nd}} b_j h_{j,c,t}^{Loss} \right)}, \forall c \in C. \quad (40)$$

where a_j and b_j are the weight of the electric and heat loads.

3. Case Studies

3.1. Case Description

The proposed strategy is evaluated using a modified P33H14 system (Figure 1) with three heat stations (HS1, HS2, and HS3) that utilize extraction-condensing CHP units and a heating boiler to supply DHS heat loads. During the pre-event phase, specific valves are typically open. The tests were conducted using Matlab R2020a on a computer with an i7-1165G7 CPU and 16 GB of memory.

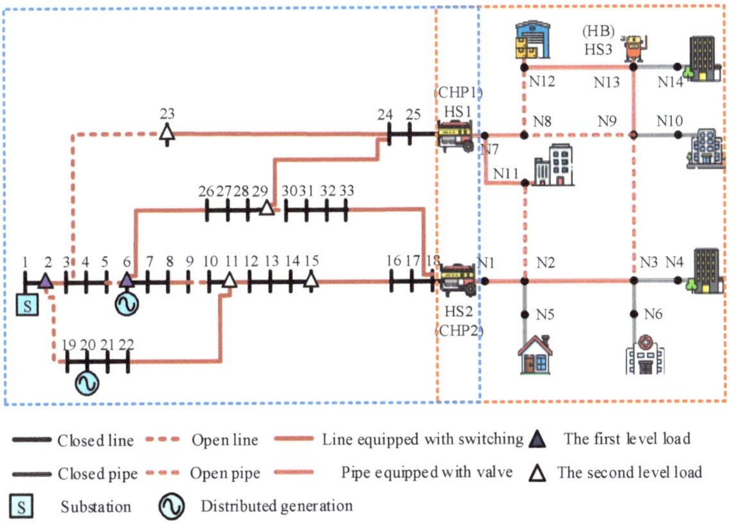

Figure 1. Structure of P33H14 system. Number represents node number, blue frame represents boundary of PDS and orange frame represents boundary of DHS.

3.2. Case Analysis

To demonstrate how the coordinated reconfiguration of PDS and DHS can improve resilience, two cases are conducted:

Case 1: Only consider PDS reconfiguration for restoration.

Case 2: Coordinated reconfiguration is considered to restore services.

3.2.1. PDS Fault Scenario

During the recovery process, several power and heat lines in the park-level Integrated Energy and Heating System (IEHS) were destroyed by natural disasters, including lines 6-26, 8-9, 11-12, and 23-24. As a result, there were major power and heat outages in the IEHS. The impact of these events is summarized in Figure 2 and Table 1 and the following conclusions can be drawn.

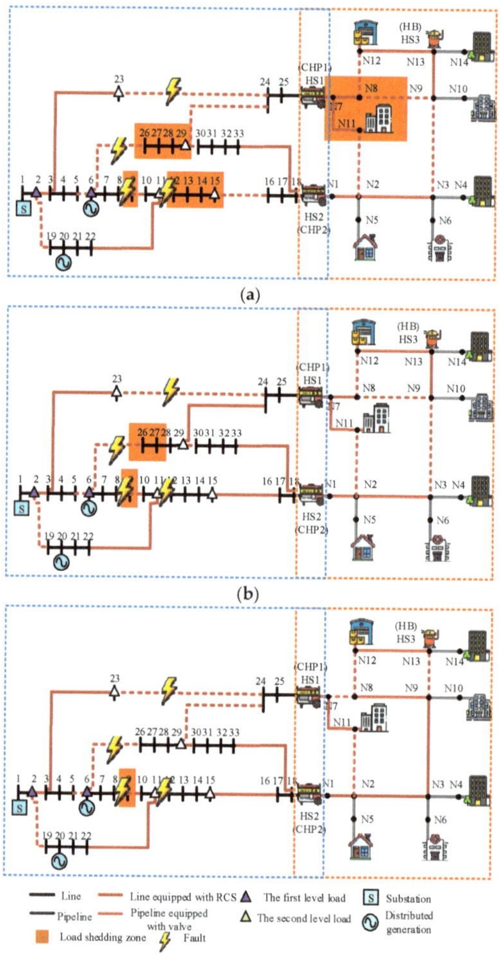

(c)

Figure 2. Topology of P33H14 during fault recovery. (**a**) Fault isolation stage; (**b**) fault restoration stage (Case 1); (**c**) fault restoration stage (Case 2). Number represents node number, blue frame represents boundary of PDS and orange frame represents boundary of DHS.

Table 1. Load curtailment and resilience metric.

Scenario	Case	Total Load Curtailment (MW)	Load Curtailment (MW) Electric	Heat	Resilience Metric
PDS fault scenario	Case 1	90.7	104.6	77.6	0.85
	Case 2	72.6	41.6	30	0.96
DHS fault scenario	Case 1	113.9	45.7	68.2	0.89
	Case 2	96.8	31.6	25.2	0.92

Firstly, the faults in the PDS propagated to DHS through CHP units, which led to the power production of CHP1 being limited during the fault isolation stage. Specifically, the load at bus 26–29 was completely lost, and the heat loads at nodes 7, 8, and 11 were partially lost.

Secondly, to improve fault repair and increase the DHS capability in natural catastrophes, the District Heating Network (DHN) reconfiguration was implemented by remotely scheduling valves and dispersing loads across heat sources. In Case 2, tie valve operation was performed on pipelines N3-N9, N7-N8, N8-N9, and N8-N12, which shifted the heat loads to CHP2. This strategy aimed to minimize DHS load shedding, and as a result, CHP1 was fully utilized.

Finally, the coordinated reconfiguration approach proved to be more effective than only PDS reconfiguration in enhancing the park-level IEHS resilience. The load curtailment decreased by 19.9%, and the value of the resilience metric increased by 12.9% in Case 2 compared to Case 1. This demonstrates the importance of a coordinated approach in improving the resilience of energy systems in the face of natural disasters.

3.2.2. DHS Fault Scenario

Table 1 displays the switch operations required for the recovery progress of the district heating system (DHS) after pipes N2-3, N2-11, and N12-13 were damaged by disasters. The results of these operations are summarized in Table 1, which leads to several conclusions.

Firstly, faults in PDS can cause simultaneous power and heat outages in DHS. During the fault isolation stage, electric load reductions at buses 30 and 31 result in partially disappearing heat loads at nodes 3, 4, and 6 and limit the power generation of CHP2.

Secondly, reconfiguring the district heating network (DHN) can improve PDS resilience and increase availability by optimally changing the topology of the heating network. Case 1 reveals that the critical heat source, CHP1′s heat output, is constrained during the isolation stage, leading to a challenging energy balance with incomplete DHS heat outage recovery despite PDS reconfiguration. In Case 2, redistributing the heat load at node 8 through valve operations in pipes N3-N9 and N8-N9 enables CHP2 operation flexibility.

Thirdly, overall load curtailment reduces by 15.0% with coordinated operation in Case 2 (Table 2).

Table 2. Switch operation during the recovery progress.

Line/Pipe	Pre-Event	Fault Isolation		Restoration	
		Case 1	Case 2	Case 1	Case 2
L3-23	0	0	0	1	1
L9-10	0	0	0	1	1
L18-33	1	1	1	0	0
L29-30	0	0	0	1	1
N3-9	0	0	1	0	1
N8-9	0	0	1	0	1

4. Conclusions

In this paper, we propose a collaborative service restoration strategy that incorporates coordinated network reconfiguration. Our approach takes into account the interaction between fault isolation and service restoration stages and emphasizes the complex coupling characteristics between the primary distribution system (PDS) and the district heating system (DHS). Through comprehensive case studies, we confirm that faults in the PDS can propagate to the DHS via coupling units. We also demonstrate that DHS reconfiguration can help expand the scope of energy supply by shifting loads among power sources. Furthermore, coordinated reconfiguration can reduce load curtailments when faults propagate between subsystems and significantly enhance the resilience of the integrated energy and heating system (IEHS) by adjusting the power production of combined heat and power (CHP) units.

For future research, we plan to consider more uncertainties, such as load fluctuations and random failures. We will also explore different ways to express similar viewpoints while ensuring the originality of the content. Overall, our study highlights the importance of coordinated network reconfiguration in enhancing the resilience of IEHSs and provides insights for future research in this area.

Author Contributions: Writing—original draft, J.W. and H.G.; writing—review & editing, J.W., H.G., Y.Y., Z.P. and H.Z.; supervision, J.W., Y.Y. and Y.L.; project administration, H.G. and Z.P.; writing—original draft, Z.P. and H.Z. All authors have read and agreed to the published version of the manuscript.

Funding: This research is funded by the Science and Technology Project of Shanxi Electric Power Company No. 52053022000K.

Conflicts of Interest: The authors declare no conflict of interest.

Nomenclature

Abbreviations	
T	Index of fault recovery periods
C	Index of fault scenarios
T_i/T_r	Index of fault isolation and service restoration periods
$k_{i,h}^{CHP}/k_{i,e}^{CHP}$	Index of CHP units i in DHS and PDS
k^{pipe}/k^{line}	Set of lines/pipes
k^{bus}/k^{bus}	Set of buses/nodes
$\pi(j)/\delta(j)$	Set of parent and child buses of bus j
$k^{CHP}/k^{HB}/k^{HS}$	Set of CHP units, heating boilers, and heat stations
S_j^{pipe-}/S_j^{pipe+}	Set of pipes flowing from/to node j
Parameters and Functions	
A_i	Power loss coefficient of SOP at bus i.
$\underline{q}_{i,c,t}^{SNOP}/\overline{q}_{i,c,t}^{SNOP}$	Minimum/maximum reactive power injections of SOP at bus i
$S_{i,c,t}^{SNOP}$	Capacity of SOP at bus i
$z_{ij,0}$	Binary variable that represents whether the line/pipe (i,j) is closed in the pre-event stage
$s_{ij,0}$	Binary variable that represents whether the line/pipe (i,j) is equipped with a switch in pre-event stage
$f_{ij,c}$	Binary variable that represents whether there is a fault on the line/pipe (i,j),
N_s	Number of heat/electric sources
N_{SOP}	Number of SOPs
N_{ij}	Number of pipes/lines
r_{ij}/x_{ij}	Binary variable that represents whether there is a fault on the line/pipe (i,j)
l_{ij}	Binary variable that represents whether there is a fault on the line/pipe (i,j)
$\overline{S}_{ij}$	ransmission capacity of the line (i,j)
$\underline{u}_j/\overline{u}_j$	Minimum/maximum square voltage magnitude of bus j.

$\underline{p}_j^{DG}/\overline{p}_j^{DG}$	Minimum/maximum power generation of DG j
$\underline{p}_j^{CHP}/\overline{p}_j^{CHP}$	Minimum/maximum power generation of CHP unit j
$\underline{v}_j/\overline{v}_j$	Minimum/maximum coefficient of power and heat generation of CHP unit j
γ_j	Coefficient between heat generation and fuel consumption of heating boiler j
$\overline{h}_{ij}^{P}$	Maximum transmission limit of the pipe (i,j)
a_j/b_j	Weight of electric and heat load j
Variables	
$p_{i,c,t}^{SNOP}/q_{i,c,t}^{SNOP}$	Active/Reactive power injection of bus i that is associated with SOP
$p_{i,c,t}^{SNOP,Loss}/q_{i,c,t}^{SNOP,Loss}$	Active/Reactive Power loss of bus i that is associated with SOP
$z_{ij,c,t}$	Binary variable that represents whether the line/pipe (i,j) is connected in the fault isolation stage during period t
$m_{i,c,t}$	Binary variable that represents whether bus i is divided into faulted regions.
$a_{ij,c,t}$	Binary variables that represent the virtual power flow between buses i and j
σ	Binary variable that represents whether SOP is in operation
$m_{i,c,t}$	Binary variable that represents whether bus i is divided into faulted regions.
$p_{j,c,t}/q_{j,c,t}$	Active/Reactive power injection of bus i that is associated with SOP
$p_{ij,c,t}/q_{ij,c,t}$	Active/Reactive power flow from bus i to bus j
$p_{j,c,t}^{DG}/q_{j,c,t}^{DG}$	Active/Reactive power generation of DG at bus i
$p_{j,c,t}^{CHP}/q_{j,c,t}^{CHP}$	Active/Reactive power generation of CHP unit at bus i
$p_{j,c,t}^{L}/q_{j,c,t}^{L}$	Electric demand of bus j
$p_{j,c,t}^{Loss}/q_{j,c,t}^{Loss}$	Load shedding of bus j
$u_{i,c,t}$	Square voltage of bus j
$h_{j,c,t}^{CHP}$	Heat generation of CHP unit j
$h_{j,c,t}^{HB}/f_{j,c,t}^{HB}$	Heat generation and fuel consumption of heating boiler j
$h_{k,c,t}^{HS}$	Heat generation of heat station k
$h_{ij,c,t}^{P,out}/h_{ij,c,t}^{P,in}/h_{ij,c,t}^{loss}$	Outlet, inlet, and loss heat quantity of pipe (i,j)

References

1. Wang, C.; Wei, W.; Wang, J.; Liu, F.; Qiu, F.; Correa-Posada, C.M.; Mei, S. Robust Defense Strategy for Gas–Electric Systems Against Malicious Attacks. *IEEE Trans. Power Syst.* **2017**, *32*, 2953–2965. [CrossRef]
2. Amirioun, M.H.; Aminifar, F.; Shahidehpour, M. Resilience-Promoting Proactive Scheduling Against Hurricanes in Multiple Energy Carrier Microgrids. *IEEE Trans. Power Syst.* **2019**, *34*, 2160–2168. [CrossRef]
3. Fotopoulou, M.; Rakopoulos, D.; Petridis, S. Decision Support System for Emergencies in Microgrids. *Sensors* **2022**, *22*, 9457. [CrossRef] [PubMed]
4. Xu, Z.; Yang, P.; Zheng, Q.; Zeng, Z. Study on Black Start Strategy of Microgrid with PV and Multiple Energy Storage Systems. In Proceedings of the 2015 18th International Conference on Electrical Machines and Systems (ICEMS), Pattaya, Thailand, 25–28 October 2015; pp. 402–408.
5. Bassey, O.; Chen, B.; Butler-Purry, K.L. Sequential Service Restoration in Distribution Systems and Microgrids Integrating Frequency Response and Varying Switching Interval. In Proceedings of the 2018 IEEE Texas Power and Energy Conference (TPEC), College Station, TX, USA, 8–9 February 2018; pp. 1–6.
6. Tan, A. In the Lower Manhattan Financial District Post-Sandy. 10. Available online: https://www.nyc.gov/html/sirr/downloads/pdf/final_report/Ch_1_SandyImpacts_FINAL_singles.pdf (accessed on 4 July 2023).
7. Manshadi, S.D.; Khodayar, M.E. Resilient Operation of Multiple Energy Carrier Microgrids. *IEEE Trans. Smart Grid* **2015**, *6*, 2283–2292. [CrossRef]
8. Yin, J.; Ren, X.; Su, S.; Yan, F.; Tao, T. Resilience-Oriented Train Rescheduling Optimization in Railway Networks: A Mixed Integer Programming Approach. *IEEE Trans. Intell. Transport. Syst.* **2023**, *24*, 4948–4961. [CrossRef]
9. Zheng, W.; Hou, Y.; Li, Z. A Dynamic Equivalent Model for District Heating Networks: Formulation, Existence and Application in Distributed Electricity-Heat Operation. *IEEE Trans. Smart Grid* **2021**, *12*, 2685–2695. [CrossRef]

10. Zheng, W.; Zhu, J.; Luo, Q. Distributed Dispatch of Integrated Electricity-Heat Systems with Variable Mass Flow. *IEEE Trans. Smart Grid* **2023**, *14*, 1907–1919. [CrossRef]
11. Zheng, W.; Lu, H.; Zhang, M. Distributed Energy Management of Multi-entity Integrated Electricity and Heat Systems: A Review of Architectures, Optimization Algorithms, and Prospects. *IEEE Trans. Smart Grid. to be published.* [CrossRef]
12. Yan, M.; He, Y.; Shahidehpour, M.; Ai, X.; Li, Z.; Wen, J. Coordinated Regional-District Operation of Integrated Energy Systems for Resilience Enhancement in Natural Disasters. *IEEE Trans. Smart Grid* **2019**, *10*, 4881–4892. [CrossRef]
13. Li, G.; Yan, K.; Zhang, R.; Jiang, T.; Li, X.; Chen, H. Resilience-Oriented Distributed Load Restoration Method for Integrated Power Distribution and Natural Gas Systems. *IEEE Trans. Sustain. Energy* **2022**, *13*, 341–352. [CrossRef]
14. Lin, Y.; Chen, B.; Wang, J.; Bie, Z. A Combined Repair Crew Dispatch Problem for Resilient Electric and Natural Gas System Considering Reconfiguration and DG Islanding. *IEEE Trans. Power Syst.* **2019**, *34*, 2755–2767. [CrossRef]
15. Pan, Z.; Guo, Q.; Sun, H. Feasible Region Method Based Integrated Heat and Electricity Dispatch Considering Building Thermal Inertia. *Appl. Energy* **2017**, *192*, 395–407. [CrossRef]
16. Cao, Y.; Wei, W.; Wu, L.; Mei, S.; Shahidehpour, M.; Li, Z. Decentralized Operation of Interdependent Power Distribution Network and District Heating Network: A Market-Driven Approach. *IEEE Trans. Smart Grid* **2019**, *10*, 5374–5385. [CrossRef]
17. Khatibi, M.; Bendtsen, J.D.; Stoustrup, J.; Molbak, T. Exploiting Power-to-Heat Assets in District Heating Networks to Regulate Electric Power Network. *IEEE Trans. Smart Grid* **2021**, *12*, 2048–2059. [CrossRef]
18. Ding, Y.; Shao, C.; Hu, B.; Bao, M.; Niu, T.; Xie, K.; Singh, C. Operational Reliability Assessment of Integrated Heat and Electricity Systems Considering the Load Uncertainties. *IEEE Trans. Smart Grid* **2021**, *12*, 3928–3939. [CrossRef]
19. Liu, Y.; Su, Y.; Xiang, Y.; Liu, J.; Wang, L.; Xu, W. Operational Reliability Assessment for Gas-Electric Integrated Distribution Feeders. *IEEE Trans. Smart Grid* **2019**, *10*, 1091–1100. [CrossRef]
20. Xue, Y.; Li, Z.; Lin, C.; Guo, Q.; Sun, H. Coordinated Dispatch of Integrated Electric and District Heating Systems Using Heterogeneous Decomposition. *IEEE Trans. Sustain. Energy* **2020**, *11*, 1495–1507. [CrossRef]
21. Liu, J.; Qin, C.; Yu, Y. Enhancing distribution system resilience with proactive islanding and RCS-based fast fault isolation and service restoration. *IEEE Trans. Smart Grid* **2020**, *11*, 2381–2395. [CrossRef]
22. Yang, Y.; Li, Z.; Mandapaka, P.V.; Lo, E.Y. Risk-averse restoration of coupled power and water systems with small pumped-hydro storage and stochastic rooftop renewables. *Appl. Energy* **2023**, *339*, 120953. [CrossRef]
23. Li, Z.; Xu, Y.; Wang, P.; Xiao, G. Coordinated preparation and recovery of a post-disaster Multi-energy distribution system considering thermal inertia and diverse uncertainties. *Appl. Energy* **2023**, *336*, 120736. [CrossRef]
24. Wang, K.; Xue, Y.; Guo, Q.; Shahidehpour, M.; Zhou, Q.; Wang, B.; Sun, H. A Coordinated Reconfiguration Strategy for Multi-Stage Resilience Enhancement in Integrated Power Distribution and Heating Networks. *IEEE Trans. Smart Grid* **2023**, *14*, 2709–2722. [CrossRef]
25. Zhang, G.; Zhang, F.; Zhang, X.; Meng, K.; Dong, Z.Y. Sequential Disaster Recovery Model for Distribution Systems With Co-Optimization of Maintenance and Restoration Crew Dispatch. *IEEE Trans. Smart Grid* **2020**, *11*, 4700–4713. [CrossRef]
26. Sarantakos, I.; Zografou-Barredo, N.-M.; Huo, D.; Greenwood, D. A Reliability-Based Method to Quantify the Capacity Value of Soft Open Points in Distribution Networks. *IEEE Trans. Power Syst.* **2021**, *36*, 5032–5043. [CrossRef]
27. Xue, Y.; Shahidehpour, M.; Pan, Z.; Wang, B.; Zhou, Q.; Guo, Q.; Sun, H. Reconfiguration of District Heating Network for Operational Flexibility Enhancement in Power System Unit Commitment. *IEEE Trans. Sustain. Energy* **2021**, *12*, 1161–1173. [CrossRef]
28. Lei, S.; Chen, C.; Song, Y.; Hou, Y. Radiality constraints for resilient reconfiguration of distribution systems: Formulation and application to microgrid formation. *IEEE Trans. Smart Grid* **2020**, *9*, 3944–3956. [CrossRef]
29. Lei, S.; Wang, J.; Hou, Y. Remote-controlled switch allocation enabling prompt restoration of distribution systems. *IEEE Trans. Power Syst.* **2018**, *5*, 3129–3142. [CrossRef]
30. Grebenyuk, G.G.; Krygin, A.A. Algorithms for optimization of the number of switchings in heat supply networks reconfiguration. *Autom Remote Control* **2007**, *12*, 2187–2197. [CrossRef]
31. Arif, A.; Cui, B.; Wang, Z. Switching Device-Cognizant Sequential Distribution System Restoration. *IEEE Trans. Power Syst.* **2022**, *37*, 317–329. [CrossRef]
32. Cheng, R.; Wang, Z.; Guo, Y. An Online Feedback-Based Linearized Power Flow Model for Unbalanced Distribution Networks. *IEEE Trans. Power Syst.* **2022**, *9*, 3552–3565. [CrossRef]
33. Lv, C.; Liang, R.; Jin, W.; Chai, Y.; Yang, T. Multi-stage resilience scheduling of electricity-gas integrated energy system with multi-level decentralized reserve. *Appl. Energy* **2022**, *7*, 119165. [CrossRef]
34. Zhu, D.; Yang, B.; Liu, Y.; Wang, Z.; Ma, K.; Guan, X. Energy management based on multi-agent deep reinforcement learning for a multi-energy industrial park. *Appl. Energy* **2022**, *4*, 118636. [CrossRef]
35. Garces, A. A linear three-phase load flow for power distribution systems. *IEEE Trans. Power Syst.* **2016**, *31*, 827–828. [CrossRef]

electronics

MDPI

Article

A Robust Interval Optimization Method for Combined Heat and Power Dispatch

Jinhao Wang [1], Junliu Zhang [1], Zhaoguang Pan [2,*], Huaichang Ge [2], Xiao Chang [1], Bin Wang [2], Shengwen Li [1] and Haotian Zhao [2]

[1] Electric Power Research Institute of State Grid Shanxi Electric Power Company, Taiyuan 030000, China; 7wjh@163.com (J.W.); zhangjunliu@sx.sgcc.com.cn (J.Z.); changxiao@sx.sgcc.com.cn (X.C.); lsw242@126.com (S.L.)
[2] Department of Electrical Engineering, Tsinghua University, Beijing 100084, China; 13261454982@163.com (H.G.); wb1984@tsinghua.edu.cn (B.W.); zhaohaotian@tsinghua.edu.cn (H.Z.)
* Correspondence: panzg09@163.com

Abstract: The increasing penetration of renewable energy, particularly wind power, and the integration of different energy systems have become two major trends in the development of energy systems. In this context, this paper proposes a robust interval optimization method for combined heat and power dispatch (CHPD) to address the challenges associated with wind power accommodation. To enhance the flexibility of a power system and support the integration of wind power, flexibility resources from a district heating system are introduced in the economic dispatch. To ensure the safety and reliability of the CHPD results, a robust interval optimization method is employed. By considering a range of possible wind power outputs, the robust interval optimization method provides a robust and reliable dispatch plan that can accommodate uncertainties and fluctuations in wind power generation. To verify the effectiveness of the proposed model and method, case studies were conducted on a 6-bus electrical power system connected with a 6-node district heating system. The results demonstrate that the proposed approach can effectively enhance the integration of wind power and improve the overall reliability and flexibility of the energy system.

Keywords: combined heat and power dispatch; robust interval optimization; wind power

Citation: Wang, J.; Zhang, J.; Pan, Z.; Ge, H.; Chang, X.; Wang, B.; Li, S.; Zhao, H. A Robust Interval Optimization Method for Combined Heat and Power Dispatch. *Electronics* **2023**, *12*, 3706. https://doi.org/10.3390/electronics12173706

Academic Editor: Adel M. Sharaf

Received: 26 July 2023
Revised: 17 August 2023
Accepted: 25 August 2023
Published: 1 September 2023

1. Introduction

In recent years, there has been a growing demand for a cleaner and more sustainable energy system. This shift in global energy priorities has led to the rapid growth of wind power as one of the key sources of renewable energy. Wind power has the ability to harness the natural power of wind and convert it into electricity, making it an attractive option for reducing greenhouse gas emissions and transitioning to a cleaner power system.

However, the integration of wind power into the existing power grid presents certain challenges. One of the main challenges is the random and fluctuating nature of wind power output. Unlike traditional power plants, which have a steady and predictable output, wind turbines are highly dependent on weather conditions and can vary in their output from one moment to the next. This poses a serious risk to the reliable operation of power systems, as the sudden changes in wind power output can destabilize the grid and lead to power outages.

To address this issue, researchers have been exploring the coordinated operation of an electrical power system (EPS) and district heating systems (DHSs) as a means of increasing the flexibility of a power system and mitigating the operational risks caused by wind power uncertainty. By coupling an EPS and DHSs through combined heat and power dispatch (CHPD), which involves the use of CHP units, boilers, pumps, and other components, it is possible to optimize the operation of a system and enhance its ability to accommodate wind power fluctuations.

Researchers have conducted numerous studies on combined heat and power dispatch (CHPD). Ref. [1] first proposed the CHPD concept, optimized the operational cost, and enhanced the wind power consumption capacity of a system by using the pipeline heat storage capacity. Ref. [2] analyzed the impact of boilers and heat pumps on wind penetration. Refs. [3,4] simulated the phase change process of heat storage, and introduced the CHP engine heat transfer process into CHPD. Ref. [5] introduced electric boilers with heat storage tanks in CHPD and found that they facilitate wind power accommodation. Refs. [6–10] investigated the thermal inertia of buildings and their power load shifting capacity. These studies have focused on optimizing the operational costs and enhancing the wind power consumption capacity of systems by using boilers, heat pumps, and the thermal inertia of buildings.

However, most of these studies have treated the CHPD problem as a deterministic one, without directly considering the uncertainty of wind power output. This means that the reliability of EPS operation cannot be guaranteed, and the results obtained are likely to be infeasible even for small wind power output perturbations. Specifically, the results may violate the power system reserve capacity limit or line transmission capacity limit. Thus, a "safe" CHPD method is required to cope with the wind power output uncertainty.

To address this limitation, researchers have started exploring stochastic optimization and chance-constrained optimization techniques to account for wind power output uncertainty in CHPD [7,11]. However, these approaches have their own limitations, such as the computational difficulty of stochastic optimization and the need for a priori knowledge of the probability distribution of uncertain parameters.

In recent years, robust optimization (RO) has emerged as a promising approach for dealing with uncertainty in CHPD. Unlike stochastic optimization and chance-constrained optimization, RO does not require explicit knowledge of the probability distribution of uncertain parameters. Instead, it focuses on finding solutions that are robust against a range of possible scenarios. Ref. [11] proposed a robust combined heat and power dispatch (R-CHPD) model that considers demand response uncertainty. Ref. [12] proposed a robust short-term CHPD scheme. Ref. [13] proposed a data-driven R-CHPD model solved by the C&CG algorithm.

However, the existing R-CHPD is formulated with a min–max or max–min objective function, which assumes that wind power can be fully absorbed even under extreme and unstable wind power conditions. In reality, wind curtailment is sometimes necessary to ensure the stability and reliability of the power system, especially during periods of low load demand. To address this issue, this paper proposes a novel robust CHPD framework that takes into account the uncertainty of wind power output and allows for wind curtailment when necessary. The proposed framework utilizes a robust interval approach, where the predicted wind output intervals are uploaded to the dispatch center. The CHPD problem is then optimized to calculate the allowable wind output within these intervals. The allowable interval is sent back to the wind farms as their control target, and CHPD for thermal units is applied accordingly. This approach ensures that the power system operates within safe limits and can effectively handle the uncertainty of wind power output.

To demonstrate the effectiveness of the proposed framework, numerical simulations are conducted in a case study. The results show that the proposed method significantly improves the reliability and stability of the power system, even under uncertain wind power conditions. The flexibility provided by the coordinated operation of the EPS and DHSs through CHPD allows for better integration of wind power and increases the overall efficiency of the system.

The paper is organized as follows. In Section 2, the paper provides a deterministic model of CHPD. This model aims to optimize the dispatch of both heat and power generation units in a system. It takes into account various constraints such as energy demand, fuel cost, and operational limitations of the units. Moving on to Section 3, a robust interval model of CHPD is presented along with its corresponding solution method. This model considers uncertainties and variations in input parameters to account for real-world scenarios where exact values are not always available. In addition, the solution method proposed in Section 3 effectively handles the interval constraints and provides practical solutions. To evaluate the performance and effectiveness of the proposed models, Section 4 presents a series of numerical simulations. The results obtained from the simulations are analyzed and compared to demonstrate the advantages and limitations of each model. Finally, in Section 5, the conclusion of the paper is provided. This section summarizes the key findings and contributions of the research. It also discusses the implications of the results and suggests potential directions for future work.

2. Deterministic Model of CHPD

2.1. Objective Function

In order to optimize the operation of units, the objective function of CHPD is designed to minimize the total operational cost. This cost is modeled as a quadratic equation, taking into account various factors that contribute to the overall cost. First, the fuel cost of non-CHP units $\sum_{t\in T} C_t^G$ is considered in the objective function. These units are responsible for generating electricity without the simultaneous production of heat. The cost of fuel consumed by these units directly impacts the overall operation cost and therefore needs to be minimized. Second, the penalty cost of wind curtailment $\sum_{t\in T} C_t^W$ is also included in the objective function. Wind power generation is inherently uncertain due to the variability and intermittency of wind resources. In situations where the wind power output exceeds the demand or the capacity of the power grid, curtailment is necessary to maintain grid stability. However, curtailment leads to a loss in potential renewable energy generation and thus incurs a penalty cost. By considering this penalty cost, the objective function aims to optimize the use of wind power while minimizing curtailment. Third, the fuel cost of CHP units $\sum_{t\in T} C_t^{CHP}$ is accounted for in the objective function. CHP units generate both electricity and heat simultaneously, making them more efficient compared to the separate generation of electricity and heat. The fuel cost associated with operating these CHP units is an important component of the total operational cost and needs to be minimized.

$$min \sum\nolimits_{t\in T} C_t^G + \sum\nolimits_{t\in T} C_t^W + \sum\nolimits_{t\in T} C_t^{CHP} \tag{1}$$

$$C_{i,t}^G = a_{0,i}^G + a_{1,i}^G P_{i,t}^G + a_{2,i}^G (P_{i,t}^G)^2, \forall i \in \Omega_G, \forall t \in T \tag{2}$$

$$C_{i,t}^W = \sigma^W (P_i^W - P_{i,t}^W)^2, \forall i \in \Omega_W, \forall t \in T \tag{3}$$

$$\begin{aligned} C_{i,t}^{CHP} = {} & a_{0,i}^{CHP} + a_{1,i}^{CHP} P_{i,t}^{CHP} + a_{2,i}^{CHP} H_{i,t}^{CHP} + a_{3,i}^{CHP} P_{i,t}^{CHP} \\ & + a_{4,i}^{CHP} (P_{i,t}^{CHP})^2 + a_{5,i}^{CHP} (H_{i,t}^{CHP})^2, \forall i \in \Omega_{CHP}, \forall t \in T \end{aligned} \tag{4}$$

where $P_{i,t}^G$ is the variable for the power output of non-CHP unit i at period t. $P_{i,t}^W$ is the variable for the power output of wind farm i during period t. $P_{i,t}^{CHP}$ and $H_{i,t}^{CHP}$ are variables for the power and heat output of CHP units i during period t, respectively. Ω_G, Ω_W, and Ω_{CHP} are sets for non-CHP units, wind farms, and CHP units, respectively. $a_{0,i}^G$, $a_{1,i}^G$, and $a_{2,i}^G$ are cost coefficients of non-CHP i. $a_{0,i}^{CHP}$, $a_{1,i}^{CHP}$, $a_{2,i}^{CHP}$, $a_{3,i}^{CHP}$, $a_{4,i}^{CHP}$, and $a_{5,i}^{CHP}$ are the cost coefficients of CHP i. σ^W is the penalty factor of wind curtailment.

2.2. DHS Constraints

Heat stations are the primary sources of heat in a DHS. The heat generated in the heat stations is transferred to the heat loads through a network of pipelines. These pipelines are designed to transport hot water or steam to various buildings and facilities within the district. The circulating water in the pipelines acts as a medium to transfer the heat from the heat stations to the heat loads. In this part, the heat stations, pipelines, and heat loads are formulated as given below.

2.2.1. Heat Station

In addition to CHP units, there are other heat sources commonly utilized in heat stations such as heating boilers and heat tanks. However, for the sake of simplicity, this discussion will focus solely on the use of CHP units as heat sources.

In engineering practice, there are two commonly employed types of CHP units: back-pressure turbines and extraction-condensing turbines. These units play a crucial role in generating heat for various applications. In order to accurately model the outputs of these CHP units, it is common practice to use convex combinations of the operational region poles [14]. By considering the different operating conditions and parameters, engineers can develop mathematical models that accurately represent the performance of these units. This modeling approach allows for better control and optimization of the CHP system, leading to improved efficiency and overall performance.

$$P_{i,t}^{CHP} = \sum_{j \in OR^i} \lambda_{i,t,j} p_{i,j}^{CHP}, H_{i,t}^{CHP} = \sum_{j \in OR^i} \lambda_{i,t,j} h_{i,j}^{CHP}, \forall t \in T, \forall i \in \Omega_{CHP} \tag{5}$$

$$0 \leq \lambda_{i,t,j}^{CHP} \leq 1, \sum_{j \in OR^i} \lambda_{i,t,j}^{CHP} = 1, \forall j \in OR^i, \forall t \in T, \forall i \in \Omega_{CHP} \tag{6}$$

where $p_{i,j}^{CHP}$ and $h_{i,j}^{CHP}$ are parameters for the power and heat output at the j-th extreme point in the operating region of CHP units i. $\lambda_{i,t,j}$ is variable for the operating point of CHP units i during period t. OR^i is a set of extreme points in the operating region of CHP units i. The heat output of CHP units is utilized for heating water:

$$H_{i,t}^{CHP} = cm_j^{HS} \left(T_{n,t}^{NS} - T_{n,t}^{NR} \right), \forall i \in \Omega_{CHP}, \forall j \in \Omega_{HS}, \forall n = N_j^{HS}, \forall t \in T \tag{7}$$

where c is a parameter for the specific thermal capacity of water. m_j^{HS} is a parameter for mass flow at heat station j. $T_{n,t}^{NS}$ and $T_{n,t}^{NR}$ are variables for the water temperature in supply and return pipes at node n during period t. Ω_{HS} is a set of heat stations. N_j^{HS} is an index of heat nodes connecting to heat station j.

2.2.2. District Heating Network

In the model of a DHN, temperature mixing Equations (8) and (9) are used to describe the process of mixing hot water from different sources. These equations take into account the flow rates and temperatures of the incoming water streams, as well as the mixed node temperature.

$$\sum_{j \in \Omega_{P+}} m_j^{PS} T_{n,t}^{NS} = \sum_{j \in \Omega_{P-}} \left(m_j^{PS} \cdot T_{j,t}^{PS,out} \right), \forall n \in \Omega_{ND}, \forall t \in T \tag{8}$$

$$\sum_{j \in \Omega_{P+}} m_j^{PR} T_{n,t}^{NR} = \sum_{j \in \Omega_{P-}} \left(m_j^{PR} \cdot T_{j,t}^{PR,out} \right), \forall n \in \Omega_{ND}, \forall t \in T \tag{9}$$

where $T_{j,t}^{PS,out}$ and $T_{j,t}^{PR,out}$ are variables representing the outflow temperatures of supply and return pipes j during period t, respectively. Ω_{ND} is a set of heat nodes. The heat loss Equations (10)–(14) are included in the model to account for the energy losses that occur during the transportation of hot water through the pipes of the DHN. These equations consider factors such as the insulation properties of the pipes, the ambient temperature, and the length and diameter of the pipes.

$$T_{j,t}^{PS,out*} = \left(1 - \left\lceil \phi_j^{PS} \right\rceil + \phi_j^{PS}\right) T_{j,t-\lceil \phi_j^{PS} \rceil}^{PS,in} + \left(\left\lceil \phi_j^{PS} \right\rceil - \phi_j^{PS} \right) T_{j,t-\lceil \phi_j^{PS} \rceil + 1}^{PS,in}, \forall j \in \Omega_n^P, \forall t \in T \tag{10}$$

$$T_{j,t}^{PS,out*} = \left(1 - \left\lceil \phi_j^{PR} \right\rceil + \phi_j^{PR}\right) T_{j,t-\lceil \phi_j^{PR} \rceil}^{PR,in} + \left(\left\lceil \phi_j^{PR} \right\rceil - \phi_j^{PR} \right) T_{j,t-\lceil \phi_j^{PR} \rceil + 1}^{PR,in}, \forall j \in \Omega_n^P, \forall t \in T \tag{11}$$

$$T_{j,t}^{PS,out} = T_t^{AM} + \left(T_{j,t}^{PS,out*} - T_t^{AM}\right) \cdot exp\left[-\frac{\lambda_j}{A_j \rho c}\left(\left\lceil \phi_j^{PS} \right\rceil - 0.5\right)\right], \forall j \in \Omega_n^P, \forall t \in T \quad (12)$$

$$T_{j,t}^{PR,out} = T_t^{AM} + \left(T_{j,t}^{PR,out*} - T_t^{AM}\right) \cdot exp\left[-\frac{\lambda_j}{A_j \rho c}\left(\left\lceil \phi_j^{PR} \right\rceil - 0.5\right)\right], \forall j \in \Omega_n^P, \forall t \in T \quad (13)$$

$$\phi_j^{PS} = \frac{\rho A_j L_j}{m_j^{PS}}, \phi_j^{PR} = \frac{\rho A_j L_j}{m_j^{PR}}, \forall j \in \Omega_n^P, \forall t \in T \quad (14)$$

where $T_{j,t}^{PS,in}$ and $T_{j,t}^{PR,in}$ are variables representing the inflow temperatures of supply and return pipes j during period t, respectively. $T_{j,t}^{PS,out*}$ and $T_{j,t}^{PR,out*}$ are the auxiliary variables indicating the pipe outlet temperature ignoring the heat loss of the pipe. Ω_n^P is a set of pipelines connecting with node n. Ω_n^{P+} and Ω_n^{P-} are sets of pipelines starting and ending at node n, respectively. T_t^{AM} is a parameter of ambient temperature at period t. ϕ_j^{PS} and ϕ_j^{PR} are the water transfer times in supply and return pipelines j. λ_j, L_j, and A_j are parameters for heat transfer, length, and cross-sectional area of pipe j, respectively. ρ is the parameter for water density. The symbol $\lceil \, \rceil$ is rounded up. To calculate the inlet temperature at each node in the DHN, Equations (15) and (16) are utilized.

$$T_{n,t}^{NR} = T_{j,t}^{PR,in}, \forall n \in \Omega_{ND}, \forall j = \Omega_n^{P+}, \forall t \in T \quad (15)$$

$$T_{n,t}^{NS} = T_{j,t}^{PS,in}, \forall n \in \Omega_{ND}, \forall j = \Omega_n^{P-}, \forall t \in T \quad (16)$$

2.2.3. Heat Loads

The heat load is presented as

$$H_{i,t}^{LD} = cm_i^{LD}\left(T_{n,t}^{NS} - T_{n,t}^{NR}\right), \forall i \in \Omega_{LD}, \forall n = N_i^{LD}, \forall t \in T \quad (17)$$

where $H_{i,t}^{LD}$ is the variable for heat load i at period t. m_i^{LD} is the parameter for the mass flow of heat load i. Ω_{LD} is a set of heat loads. N_i^{LD} is an index of heat nodes linked to heat load j. To guarantee the heating demand, the temperature of the heat load is satisfied:

$$\underline{T}_n^{NR} \le T_{n,t}^{NR} \le \overline{T}_n^{NR}, \forall i \in \Omega_{LD}, \forall n = N_i^{LD}, \forall t \in T \quad (18)$$

where $\underline{T}_n^{NR}$ and $\overline{T}_n^{NR}$ are the water temperature boundaries at node n.

2.3. EPS Constraints

Here, the DC model is used for EPS. The power balance constraints (19) ensure that the total power generation in the EPS matches the total power demand. This constraint is essential for maintaining system stability and avoiding power shortages or overloads. Transmission capacity constraints (20) are imposed to limit the amount of power that can flow through the transmission lines. These constraints take into account the capacity limitations of the transmission infrastructure and prevent congestion or overloading. The thermal unit ramping constraints (21) and (22) specify the rate at which thermal units can change their generation output. These constraints ensure that the units' output changes smoothly and gradually, preventing sudden and drastic changes that could destabilize the system. The thermal units' generation output constraints (23) and (24) define the range within which the generation output of thermal units must lie. These constraints are based on the characteristics and limitations of each thermal unit, such as its maximum and minimum output levels. Wind farms' power output constraints (25) are used to limit the power output of wind farms. Since wind power is intermittent and dependent on weather conditions, these constraints ensure that the wind farms' output remains within certain bounds, preventing excessive reliance on unreliable wind power. Finally, the spinning reserve constraints (26)–(28) are imposed to ensure that there is sufficient reserve capacity in the system to handle unexpected changes in demand or generation. In this study, the spinning reserve is assumed to be provided by non-CHP units, as the spinning reserve ability of CHP units is limited by their heat loads.

$$\sum_{i\in\Omega_G} P_{i,t}^{G} + \sum_{i\in\Omega_{CHP}} P_{i,t}^{CHP} + \sum_{i\in\Omega_W} P_{i,t}^{W} = \sum_{i\in\Omega_{bus}} P_{i,t}^{LD}, \forall t \in T \tag{19}$$

$$\left|\sum_{j\in\Omega_{bus}} SF_{l,j}\left(\sum_{i\in S_j^G} P_{i,t}^{G} + \sum_{i\in S_j^{CHP}} P_{i,t}^{CHP} + \sum_{i\in S_j^W} P_{i,t}^{W} - P_{j,t}^{LD}\right)\right| \le F_l, \forall l \in \Omega_{line}, \forall t \in T \tag{20}$$

$$-RD_i \cdot \Delta t \le P_{i,t}^{G} - P_{i,t-1}^{G} \le -RU_i \cdot \Delta t, \forall i \in \Omega_G, \forall t \in T \tag{21}$$

$$-RD_i \cdot \Delta t \le P_{i,t}^{CHP} - P_{i,t-1}^{CHP} \le -RU_i \cdot \Delta t, \forall i \in \Omega_{CHP}, \forall t \in T \tag{22}$$

$$\underline{P}_i^{G} \le P_{i,t}^{G} \le \overline{P}_i^{G}, \forall i \in \Omega_G, \forall t \in T \tag{23}$$

$$\underline{P}_i^{CHP} \le P_{i,t}^{CHP} \le \overline{P}_i^{CHP}, \forall i \in \Omega_{CHP}, \forall t \in T \tag{24}$$

$$0 \le P_{i,t}^{W} \le \overline{P}_i^{W}, \forall i \in \Omega_W, \forall t \in T \tag{25}$$

$$0 \le ru_{i,t} \le RU_i, ru_{i,t} \le \overline{P}_i^{G} - P_{i,t}^{G}, \forall i \in \Omega_G, \forall t \in T \tag{26}$$

$$0 \le rd_{i,t} \le RD_i, rd_{i,t} \le P_{i,t}^{G} - \underline{P}_i^{G}, \forall i \in \Omega_G, \forall t \in T \tag{27}$$

$$\sum_{i\in\Omega_G} ru_{i,t} \ge SU, \sum_{i\in\Omega_G} rd_{i,t} \ge SD, \forall t \in T \tag{28}$$

where $ru_{i,t}$ and $rd_{i,t}$ are variables for the upward and downward spinning reserve of non-CHP unit i at period t. $SF_{i,j}$ is the shift factor of bus j to line l. F_l is the maximum transmission flow of line l. RU_i and RD_i are parameters for the upward ramping capacity and downward ramping capacity of units i, respectively. $P_{i,t}^{LD}$ is a parameter for electric load connecting to bus i during period t. $\underline{P}_i^{G}$ and $\overline{P}_i^{G}$ are parameters for the power output boundaries of non-CHP unit i, respectively. $\underline{P}_i^{CHP}$ and $\overline{P}_i^{CHP}$ are parameters for the power output boundaries of CHP unit i. $\overline{P}_i^{W}$ is a parameter for the forecast output of wind farm i. SU and SD are parameters for system-wide upward and downward ramping capacity. Ω_{bus} and Ω_{line} are sets for bus and line, respectively. S_j^G, S_j^{CHP}, and S_j^W are sets for non-CHP units, CHP units, and wind farms connected with bus j, respectively.

3. Robust Model of CHPD and the Solution Method

Considering that the CHPD model proposed in Section 2 is unable to capture wind power uncertainty, this section first proposes a robust CHPD model, and then the robust interval optimization method is used to handle the R-CHPD problem.

3.1. Formulation of Robust CHPD

In the R-CHPD framework, the uncertain nature of wind power output is taken into account by representing it as an interval $\left[\underline{p}^W_{i,t}, \overline{p}^W_{i,t}\right]$ rather than a single value $\overline{P}^W_{i,t}$. This interval captures the potential range of wind power generation, considering both the lower bound $\underline{p}^W_{i,t}$ and upper bound $\overline{p}^W_{i,t}$. By considering this range, operators can analyze the worst-case scenarios and develop robust strategies to ensure the system's safe and reliable operation. Typically, there are four worst-case scenarios, as follows:

1. The worst-case scenario for the upward-spinning reserve constraint is

$$R_t^U = \left\{ \begin{array}{l} \min\limits_{P^W_{i,t}} \left(\sum_{i \in \Omega_G} P^G_{i,t} + \sum_{i \in \Omega_{CHP}} P^{CHP}_{i,t} + \sum_{i \in \Omega_W} P^W_{i,t} + \sum_{i \in \Omega_G} ru_{i,t} - \sum_{i \in \Omega_{bus}} P^{LD}_{i,t} \right) \geq 0 \\ s.t. \tilde{\underline{p}}^W_{i,t} \leq P^W_{i,t} \leq \tilde{\overline{p}}^W_{i,t}, \forall i \in \Omega_W, \forall t \in T \end{array} \right\} \tag{29}$$

where $\tilde{\underline{p}}^W_{i,t}$ and $\tilde{\overline{p}}^W_{i,t}$ are the decision variables in a robust interval optimization problem which satisfies

$$\tilde{\underline{p}}^W_{i,t} \leq \underline{p}^W_{i,t}, \tilde{\overline{p}}^W_{i,t} \leq \overline{p}^W_{i,t}, \forall i \in \Omega_W, \forall t \in T \tag{30}$$

In this scenario, the system must be prepared for the maximum amount of wind power generation, which would require a surplus of resources that can be quickly dispatched to balance the excess power.

2. The worst-case scenario for the downward spinning reserve constraint is

$$R_t^D = \left\{ \begin{array}{l} \min\limits_{P^W_{i,t}} \left(\sum_{i \in \Omega_{bus}} P^{LD}_{i,t} - \sum_{i \in \Omega_G} P^G_{i,t} - \sum_{i \in \Omega_{CHP}} P^{CHP}_{i,t} - \sum_{i \in \Omega_W} P^W_{i,t} - \sum_{i \in \Omega_G} rd_{i,t} \right) \geq 0 \\ s.t. \tilde{p}^W_{i,t} \leq P^W_{i,t} \leq \tilde{\overline{p}}^W_{i,t}, \forall i \in \Omega_W, \forall t \in T \end{array} \right\} \tag{31}$$

For these worst-case scenarios, the total upward power adjustment Δru_t and downward upward power adjustment Δrd_t are calculated as

$$\Delta ru_t = \sum\nolimits_{i \in \Omega_G} ru_{i,t} - R_t^U, \Delta rd_t = \sum\nolimits_{i \in \Omega_G} rd_{i,t} - R_t^D, \forall t \in T \tag{32}$$

$$\Delta ru_t = \sum\nolimits_{i \in \Omega_G} \Delta ru_{i,t}, \Delta rd_t = \sum\nolimits_{i \in \Omega_G} \Delta rd_{i,t}, \forall t \in T \tag{33}$$

Δru_t and Δrd_t should be shared by all the non-CHP units, so the following constraints exist:

$$0 \leq \Delta ru_{i,t} \leq ru_{i,t}, 0 \leq \Delta rd_{i,t} \leq rd_{i,t}, i \in \Omega_G, \forall t \in T \tag{34}$$

$$P^G_{i,t} + \Delta ru_{i,t} - P^G_{i,t-1} + \Delta rd_{i,t-1} \leq \Delta PU^G_{i,t}, i \in \Omega_G, \forall t \in T \tag{35}$$

$$P^G_{i,t-1} + \Delta ru_{i,t-1} - P^G_{i,t} + \Delta rd_{i,t} \leq \Delta PD^G_{i,t}, i \in \Omega_G, \forall t \in T \tag{36}$$

3. The worst-case scenario for the positive transmission interface flow constraint is

$$L_{l,t}^{U}=\left\{\begin{array}{l}\max_{P_{i,t}^{W}}\left(\sum_{j\in\Omega_{bus}}SF_{l,j}\left(\sum_{i\in S_j^G}P_{i,t}^{G}+\sum_{i\in S_j^{CHP}}P_{i,t}^{CHP}+\sum_{i\in S_j^W}P_{i,t}^{W}-P_{j,t}^{LD}\right)\right)\le F_l\\ s.t.\underset{-i,t}{\tilde{p}}^{W}\le P_{i,t}^{W}\le\tilde{p}_{i,t}^{\overline{W}},\forall i\in\Omega_W,\forall l\in\Omega_{line},\ \forall t\in T\end{array}\right\} \tag{37}$$

4. The worst-case scenario for the negative transmission interface flow constraint is

$$L_{l,t}^{D}=\left\{\begin{array}{l}\min_{P_{i,t}^{W}}\left(\sum_{j\in\Omega_{bus}}SF_{l,j}\left(\sum_{i\in S_j^G}P_{i,t}^{G}+\sum_{i\in S_j^{CHP}}P_{i,t}^{CHP}+\sum_{i\in S_j^W}P_{i,t}^{W}-P_{j,t}^{LD}\right)\right)\ge -F_l\\ s.t.\underset{-i,t}{\tilde{p}}^{W}\le P_{i,t}^{W}\le\tilde{p}_{i,t}^{\overline{W}},\forall i\in\Omega_W,\forall l\in\Omega_{line},\forall t\in T\end{array}\right\} \tag{38}$$

Moreover, the penalty cost function of possible wind curtailment for wind farms is transferred as

$$C_{i,t}^{W}=\sigma^{W}\left(\left(\overline{p}_{i,t}^{W}-\tilde{p}_{i,t}^{\overline{W}}\right)^{2}+\left(\underset{-i,t}{p}^{W}-\underset{-i,t}{\tilde{p}}^{W}\right)^{2}\right),\forall i\in\Omega_W,\forall t\in T \tag{39}$$

The detailed model of R-CHPD is summarized as follows:

$$\begin{array}{cl}\min\limits_{P_{i,t}^{G},P_{i,t}^{CHP},\underset{-i,t}{\tilde{p}}^{W},\tilde{p}_{i,t}^{\overline{W}},H_{i,t}^{CHP}} & Equation\ (1)\\ s.t. & Constraints\ (5)\text{–}(19),(21)\text{–}(24),(29)\text{–}(38)\end{array} \tag{40}$$

where $U^{\beta}\left(\underset{-i,t}{\tilde{p}}^{W},\tilde{p}_{i,t}^{\overline{W}}\right)=\left\{\tilde{p}_{i,t}^{W}\middle|\underset{-i,t}{\tilde{p}}^{W}\le\tilde{p}_{i,t}^{W}\le\tilde{p}_{i,t}^{\overline{W}}\right\}$ represents the adjustable uncertainty sets.

3.2. Model Simplification

For brevity, the R-CHPD in (40) can be written in a compact form

$$\begin{array}{lll}\min\limits_{x,\underset{-}{\tilde{y}},\bar{\tilde{y}}} & f\left(x,\underset{-}{\tilde{y}},\bar{\tilde{y}}\right) & \\ s.t. & Dx+F\tilde{y}\le c,\forall\tilde{y}\in\left[\underset{-}{\tilde{y}},\bar{\tilde{y}}\right] & (a)\\ & \underset{-}{\tilde{y}}\le\underset{-}{y},\bar{\tilde{y}}\le\bar{y} & (b)\end{array} \tag{41}$$

where x refers to variables $P_{i,t}^{G}$, $P_{i,t}^{CHP}$, $ru_{i,t}$, $rd_{i,t}$, $\Delta ru_{i,t}$, $\Delta rd_{i,t}$, R_t^U, R_t^D, $L_{l,t}^U$, $L_{l,t}^D$, $H_{i,t}^{CHP}$, $\lambda_{i,t,j}$, $T_{n,t}^{NS}$, $T_{n,t}^{NR}$, $T_{j,t}^{PS,in}$, $T_{j,t}^{PS,out}$, $T_{j,t}^{PR,in}$, and $T_{j,t}^{PR,out}$. The $\underset{-}{\tilde{y}}$ and $\bar{\tilde{y}}$ are variable vectors that represent $\underset{-i,t}{\tilde{p}}^{W}$ and $\tilde{p}_{i,t}^{\overline{W}}$, respectively. The $\tilde{y}$ is an uncertainty parameter vector that denotes the actual output of wind power $\tilde{p}_{i,t}^{W}$. $Dx+F\tilde{y}\le c$ refers to constraints (5)–(19), (21)–(24), and (29)–(38). $\underset{-}{\tilde{y}}\le\underset{-}{y},\bar{\tilde{y}}\le\bar{y}$ corresponds to constraints (30).

In (41), there is $Dx+F\tilde{y}\le c$ for $\forall\tilde{y}\in\left[\underset{-}{\tilde{y}},\bar{\tilde{y}}\right]$, which is represented without loss generality as

$$\max\left\{Dx+F\tilde{y}\middle|\tilde{y}\in\left[\underset{-}{\tilde{y}},\bar{\tilde{y}}\right]\right\}\le c \tag{42}$$

i.e.,

$$\begin{cases} \boldsymbol{D}_i \boldsymbol{x} + \max\limits_{\tilde{y}} (\boldsymbol{F}_i \tilde{\boldsymbol{y}}) \leq c_i & (a) \\ \tilde{y} \in \left[\underline{\tilde{y}}, \overline{\tilde{y}} \right] & (b) \end{cases} \tag{43}$$

where $\boldsymbol{D}_i$ and $\boldsymbol{F}_i$ is row i of the matrix $\boldsymbol{D}$ and $\boldsymbol{F}$, respectively. $\boldsymbol{c}_i$ is entry i of the vector $\boldsymbol{c}$. Equation (43) can be further transformed into

$$\begin{cases} \boldsymbol{D}_i \boldsymbol{x} + \max\limits_{\tilde{y}} \left(\boldsymbol{F}_i \left(\underline{\tilde{y}} + w \left(\overline{\tilde{y}} - \underline{\tilde{y}} \right) \right) \right) \leq c_i, \forall i & (a) \\ 0 \leq w \leq 1 & (b) \end{cases} \tag{44}$$

and the dual problem is expressed as

$$\begin{aligned} \min_{u_i} \quad & \boldsymbol{D}_i \boldsymbol{x} + \boldsymbol{F}_i \underline{\tilde{y}} + \boldsymbol{1}^T \boldsymbol{u}_i \\ s.t. \quad & \boldsymbol{u}_i \geq \boldsymbol{F}_i \left(\overline{\tilde{y}} - \underline{\tilde{y}} \right) \end{aligned}, \forall i \tag{45}$$

Based on duality theory, Equation (46) holds.

$$\boldsymbol{D}_i \boldsymbol{x} + \boldsymbol{F}_i \tilde{\boldsymbol{y}} \leq \boldsymbol{D}_i \boldsymbol{x} + \boldsymbol{F}_i \underline{\tilde{y}} + \boldsymbol{1}^T \boldsymbol{u}_i, \forall i \tag{46}$$

Consequently, (41) is equivalent to the following equivalent models:

$$\begin{aligned} \min_{x, \underline{\tilde{y}}, \overline{\tilde{y}}} \quad & f\left(x, \underline{\tilde{y}}, \overline{\tilde{y}} \right) \\ s.t. \quad & \boldsymbol{D}_i \boldsymbol{x} + \boldsymbol{F}_i \underline{\tilde{y}} + \boldsymbol{1}^T \boldsymbol{u}_i \leq \boldsymbol{c}, \forall i \\ & \boldsymbol{u}_i \geq \boldsymbol{F}_i \overline{\tilde{y}} - \boldsymbol{F}_i \underline{\tilde{y}}, \forall i \\ & \underline{\tilde{y}} \leq \underline{y}, \overline{\tilde{y}} \leq \overline{y}, \\ & \boldsymbol{u}_i \geq \boldsymbol{0}. \end{aligned} \tag{47}$$

The model presented in (47) is a quadratic programming problem, which can be handled directly. However, when the TPS and DHS are managed by different entities, it is impractical to solve (47) in a centralized manner. The use of the heterogeneous decomposition method allows for efficient and effective management of the TPS and DHS. By decomposing the problem and finding distributed solutions, the overall optimization of the R-CHPD system can be achieved. Due to the fact that distributed solution methods are not the focus of this paper, no further description of these is provided here. The detailed solution process can be found in [15].

4. Case Study

The performance of the proposed robust interval optimization method is tested using a 6-bus EPS connecting with a 6-node DHS (P6H6), as shown in Figure 1. Figure 2a shows the electric loads and heat loads, and Figure 2b shows the forecast wind power. The transmission interface capacity is 50 MW. For conciseness, other system parameters are provided in [16]. To analyze the performance of the proposed model, three cases were established:

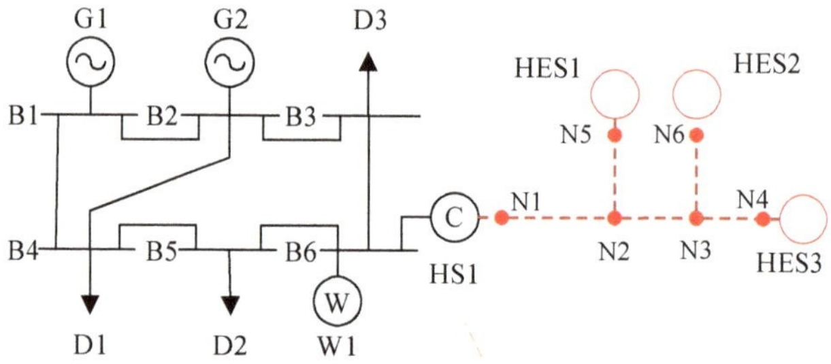

Figure 1. P6H6 system.

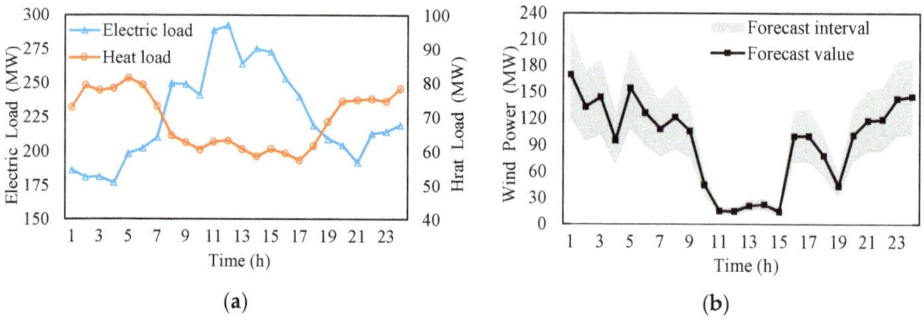

Figure 2. (**a**) Electric load and heat load; (**b**) forecast wind power.

Case 1: isolated heat and power dispatch (IHPD) mode, where the EPS and DHS are independently operated. In this mode, the DHS operator dispatches the CHP units to meet the heat demand. Subsequently, the EPS determines the unit dispatch strategies based on the constraints of CHP heat output.

Case 2: coordinated heat and power dispatch (CHPD) mode, where the EPS and DHS are dispatched in a coordinated manner. In this case, the CHP units are optimally dispatched to satisfy both the heat and power demands.

Case 3: Based on Case 2, the wind power output uncertainty is considered, and R-CHPD is performed.

The major results of the study are presented in Table 1 and Figure 3. In the case of IHPD, the heat output of CHP units is always matched to the heat demand. Due to the heat-driven nature of CHP units, the power output of these units needs to be maintained at a high level during the night, leading to significant curtailment of wind power. As shown in Table 1, the cost of wind power curtailment amounts to USD 1506.

Table 1. Test results.

Title 1	IHPD	CHPD	R-CHPD
$\sum_{t\in T} C_t^G$	USD 79,603	USD 76,548	USD 77,362
$\sum_{t\in T} C_t^{CHP}$	USD 28,329	USD 26,421	USD 26,958
$\sum_{t\in T} C_t^W$	USD 1506	USD 501	USD 1703
Total cost	USD 109,438	USD 103,470	USD 106,023

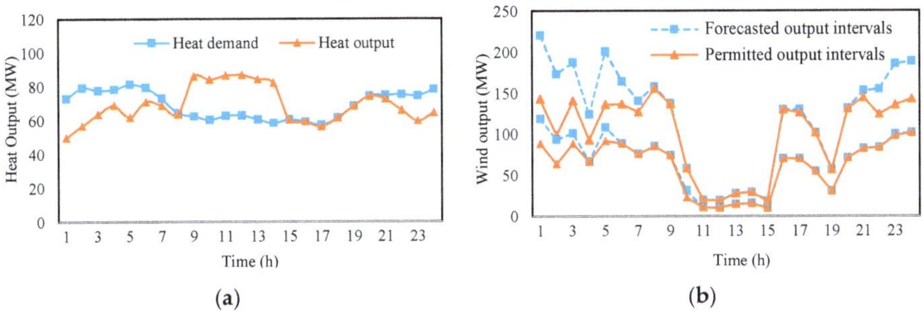

Figure 3. (**a**) Heat output in Case 2; (**b**) wind output of W1 in Case 3.

In CHPD, the heat output of the CHP unit is not strictly limited by heat loads. It utilizes the heat storage of the DHN so that the output of CHP units can be flexibly adjusted for better wind power penetration. In contrast to IHPD, the wind curtailment decreases by 3.84%, and the total cost is reduced by 5.45%.

It should be noted that in the CHPD, the wind power output is assumed to be certain, which may not be realistic in practice. In order to account for wind power uncertainties, R-CHPD requires non-CHP units to maintain a larger reserve capacity to ensure system operational security. As a result, the total cost increases by 2.47% compared to CHPD. However, it is important to highlight that the dispatch strategies in R-CHPD remain feasible as long as the wind power generation falls within the permitted output interval.

5. Conclusions

To address the issue of wind curtailment, this paper proposes a robust interval optimization technique for CHPD. Instead of using fixed numerical values as control targets for wind farms, interval values are utilized, which allows for a more flexible and adaptable approach. It is important to note that while this paper focuses on the uncertainty of wind power output, the robust interval optimization method can also be applied to other uncertainties present in the system. For instance, in real-world scenarios, there may be uncertainty in both the electric and heat loads. In such cases, these uncertain loads can be treated as negative outputs and combined with the wind power output to construct the uncertainty set.

One limitation of this method is that when the prediction error is large, the obtained results may be too conservative. In other words, the system may overly prioritize robustness at the expense of economic efficiency. To address this issue, the next step in our research is to improve the prediction interval. By refining the prediction interval, we aim to strike a balance between robustness and economy, ensuring that the system operates optimally while accounting for uncertainties. By enhancing the prediction interval, we expect to achieve a more accurate estimation of the uncertainty bounds, allowing for better decision-making in resource planning and emergency measures. This will ultimately lead to improved system performance and the effective utilization of the robust interval optimization technique in CHPD.

Author Contributions: Investigation, J.W.; Writing—review & editing, J.W., J.Z. and Z.P.; Project administration, J.W., J.Z., B.W. and S.L.; Validation J.Z. and H.G.; Writing—Original draft, Z.P., H.G. and H.Z.; Methodology, H.G.; Data curation, X.C. and B.W.; Visualization, X.C.; Supervision, S.L.; Methodology, H.Z. All authors have read and agreed to the published version of the manuscript.

Funding: This research was funded by the Science and Technology Project of Shanxi Electric Power Company no. 52053022000K.

Data Availability Statement: No new data is created.

Conflicts of Interest: The authors declare no conflict of interest.

References

1. Li, Z.; Wu, W.; Shahidehpour, M.; Wang, J.; Zhang, B. Combined Heat and Power Dispatch Considering Pipeline Energy Storage of District Heating Network. *IEEE Trans. Sustain. Energy* **2016**, *7*, 12–22. [CrossRef]
2. Chen, X.; McElroy, M.B.; Kang, C. Integrated Energy Systems for Higher Wind Penetration in China: Formulation, Implementation and Impacts. *IEEE Trans. Power Syst.* **2017**, *33*, 1309–1319. [CrossRef]
3. Dai, Y.; Chen, L.; Min, Y.; Mancarella, P.; Chen, Q.; Hao, J.; Hu, K.; Xu, F. Integrated Dispatch Model for Combined Heat and Power Plant With Phase-Change Thermal Energy Storage Considering Heat Transfer Process. *IEEE Trans. Sustain. Energy* **2018**, *9*, 1234–1243. [CrossRef]
4. Dai, Y.; Chen, L.; Min, Y.; Mancarella, P.; Chen, Q.; Hao, J.; Hu, K.; Xu, F. A General Model for Thermal Energy Storage in Combined Heat and Power Dispatch Considering Heat Transfer Constraints. *IEEE Trans. Sustain. Energy* **2018**, *9*, 1518–1528. [CrossRef]
5. Wu, C.; Gu, W.; Jiang, P.; Li, Z.; Cai, H.; Li, B. Combined Economic Dispatch Considering the Time-Delay of District Heating Network and Multi-Regional Indoor Temperature Control. *IEEE Trans. Sustain. Energy* **2018**, *9*, 118–127. [CrossRef]
6. Zheng, J.; Zhou, Z.; Zhao, J.; Wang, J. Effects of the Operation Regulation Modes of District Heating System on an Integrated Heat and Power Dispatch System for Wind Power Integration. *Appl. Energy* **2018**, *230*, 1126–1139. [CrossRef]
7. Huang, X.; Xu, Z.; Sun, Y.; Xue, Y.; Wang, Z.; Liu, Z.; Li, Z.; Ni, W. Heat and Power Load Dispatching Considering Energy Storage of District Heating System and Electric Boilers. *J. Mod. Power Syst. Clean Energy* **2018**, *6*, 992–1003. [CrossRef]
8. Wang, K.; Xue, Y.; Guo, Q.; Shahidehpour, M.; Zhou, Q.; Wang, B.; Sun, H. A Coordinated Reconfiguration Strategy for Multi-Stage Resilience Enhancement in Integrated Power Distribution and Heating Networks. *IEEE Trans. Smart Grid* **2023**, *14*, 2709–2722. [CrossRef]
9. Du, Y.; Xue, Y.; Lu, L.; Yu, C.; Zhang, J. A Bi-Level Co-Expansion Planning of Integrated Electric and Heating System Considering Demand Response. *Front. Energy Res.* **2022**, *10*, 999948. [CrossRef]
10. Li, Y.; Zou, Y.; Tan, Y.; Cao, Y.; Liu, X.; Shahidehpour, M.; Tian, S.; Bu, F. Optimal Stochastic Operation of Integrated Low-Carbon Electric Power, Natural Gas, and Heat Delivery System. *IEEE Trans. Sustain. Energy* **2018**, *9*, 273–283. [CrossRef]
11. Majidi, M.; Mohammadi-Ivatloo, B.; Anvari-Moghaddam, A. Optimal Robust Operation of Combined Heat and Power Systems with Demand Response Programs. *Appl. Therm. Eng.* **2019**, *149*, 1359–1369. [CrossRef]
12. Nazari-Heris, M.; Mohammadi-Ivatloo, B.; Gharehpetian, G.B.; Shahidehpour, M. Robust Short-Term Scheduling of Integrated Heat and Power Microgrids. *IEEE Syst. J.* **2019**, *13*, 3295–3303. [CrossRef]
13. Shui, Y.; Gao, H.; Wang, L.; Wei, Z.; Liu, J. A Data-Driven Distributionally Robust Coordinated Dispatch Model for Integrated Power and Heating Systems Considering Wind Power Uncertainties. *Int. J. Electr. Power Energy Syst.* **2019**, *104*, 255–258. [CrossRef]
14. Du, Y.; Xue, Y.; Wu, W.; Shahidehpour, M.; Shen, X.; Wang, B.; Sun, H. Coordinated Planning of Integrated Electric and Heating System Considering the Optimal Reconfiguration of District Heating Network. *IEEE Trans. Power Syst.* **2023**. early access. [CrossRef]
15. Xue, Y.; Li, Z.; Lin, C.; Guo, Q.; Sun, H. Coordinated Dispatch of Integrated Electric and District Heating Systems Using Heterogeneous Decomposition. *IEEE Trans. Sustain. Energy* **2020**, *11*, 1495–1507. [CrossRef]
16. Test Data for R-CHPD. Available online: https://docs.google.com/spreadsheets/d/1TNrEsWwrCDvDcEuHb7VnUQcul6cFcCoN/edit?usp=sharing&ouid=115924522171410221124&rtpof=true&sd=true (accessed on 26 July 2023).

electronics

MDPI

Article

Optimal Scheduling of Virtual Power Plant Considering Reconfiguration of District Heating Network

Jinhao Wang [1], Zhaoguang Pan [2,*], Shengwen Li [1], Huaichang Ge [2], Gang Yang [1] and Bin Wang [2]

1 Electric Power Research Institute of State Grid Shanxi Electirc Power Company, Taiyuan 030000, China
2 Department of Electrical Engineering, Tsinghua University, Beijing 100084, China
* Correspondence: panzg09@163.com

Abstract: A combined heat and power virtual power plant (CHP-VPP) can effectively control the distributed resources in an electric–thermal coupling system and solve the problem of lack of flexibility caused by large-scale renewable energy grid connection. Similar to the optimal reconfiguration of distribution network topology by operating switches, the district heating system is also equipped with tie and sectionalizing valves to realize the optimal adjustment of district heating network (DHN) topology, which provides an economical and effective method for improving the power system's flexibility. Based on this, this paper proposes a CHP-VPP economic scheduling model considering reconfigurable DHN. Firstly, the energy flow model is introduced to reduce the computational complexity. Secondly, adaptive robust optimization solved by the column-and-constraint generation algorithm is used to settle the randomness of wind power to ensure that the results are feasible in all worst scenarios. Finally, the feasibility of the proposed model is illustrated by case studies based on an actual CHP-VPP. The results show that compared with the reference case, considering the reconfigurability of DHN in the CHP-VPP optimization scheduling process can reduce the cost by about 2.78%.

Keywords: virtual power plants; economic scheduling; reconfigurable DHN; adjustable robust optimization

Citation: Wang, J.; Pan, Z.; Li, S.; Ge, H.; Yang, G.; Wang, B. Optimal Scheduling of Virtual Power Plant Considering Reconfiguration of District Heating Network. *Electronics* **2023**, *12*, 3409. https://doi.org/10.3390/electronics12163409

Academic Editor: J. C. Hernandez

Received: 10 July 2023
Revised: 29 July 2023
Accepted: 4 August 2023
Published: 11 August 2023

1. Introduction

The world's energy shortage and environmental pollution have become increasingly serious in recent years, and the deficiencies of traditional energy power generation have been highlighted [1,2]. Renewable energy has become the primary focus of global energy development in the future. By the end of 2021, global renewable energy generation was 3064 GW, an increase of 9.1% compared to 2020 [3]. However, the intermittence and volatility of renewable energy generation lead to potential security risks in the actual operation of the power system, which hinders the further development of renewable energy consumption capacity.

As an important distributed resource technology, virtual power plants (VPPs) provide a solution to improve renewable energy consumption [4,5]. Ref. [6] established an overall economic optimization model of multi-generator units under VPP mode, and the energy storage equipment was used to effectively reduce the rate of 'abandoning wind and light'. Ref. [7] proposed an operation mode of electric vehicle VPPs participating in the ancillary service market to promote deep peak shaving of heat power and consumption of intermittent renewable energy. Also, researchers noted that uncertainty is a significant factor affecting the operation of VPPs. Robust optimization [8], stochastic optimization [9] and opportunity constraints [10] are popularly used to handle uncertainty.

Meanwhile, with the wide application of electric–thermal coupling components like combined heat and power (CHP) units and electric boilers (EBs), the coupling relationship between electricity and heat has become closer [11,12]. The combined heat and power virtual power plant (CHP-VPP), as a coupling system of electricity and heat, has aroused the interests of researchers. Ref. [13] studied the financial benefits of heat load management in a CHP-VPP comprising a micro-CHP unit, a heat pump and residential heat storage. Ref. [14] constructed a CHP-VPP cooperative game model based on the improved Sharpley value method.

However, the above studies ignore the district heating network (DHN) model in VPP or assume that the topology of DHN is constant. Similar to the topology reconfiguration of the power distribution network (PDN) by controlling remote switches, the topology of DHN in CHP-VPP can also be reconfigured by distance control pipeline valves [15]. On the one hand, the heat load after reconfiguration can be redistributed between heat sources to obtain a more economical scheduling plan to decrease the cost of the whole system. On the other hand, the operation flexibility of CHP units is enhanced after reconfiguration, and the capacity of renewable energy consumption is also improved. Some researchers have studied the application of reconfigurable DHN in integrated energy systems. Ref. [16] studied the seasonal DHN reconstruction to improve the rationality of the planning strategy. Ref. [17] considered the collaborative reconfiguration of the power distribution system (PDS) and district heating system (DHS) to enhance the flexibility of park-level integrated power and heating systems.

Accordingly, this paper takes a CHP-VPP with wind turbines (WTs), conventional generators, CHP units and EBs as an example to establish an economic dispatch model considering reconfigurable DHN. The following are the contributions of this paper:

(1) For the first time, the reconfigurable DHN is considered in the economic scheduling problem of CHP-VPP, which can significantly reduce the operating cost of CHP-VPP. The energy flow model is introduced to reduce the difficulty of solving the model.
(2) Adaptive robust optimization (ARO) solved by the column-and-constraint generation (C&CG) algorithm is applied to settle the randomness of wind power, and in order to avoid too conservative decision-making, a robust control coefficient is introduced for adjustment.
(3) The effectiveness and versatility of the CHP-VPP economic dispatch model considering reconfigurable DHN are verified by an example analysis based on an actual CHP-VPP in China.

2. Optimal Scheduling Model of CHP-VPP Considering Reconfigurable DHN

Figure 1 presents the internal framework of the CHP-VPP. The electric power is generated by CHP units, WTs and conventional generators. In addition, CHP-VPP can interact with the power grid through tie lines and provide electricity through the PDN. The heat in CHP-VPP is generated by the heating station with a heat source (CHP units and EBs) and transported by the circulating hot water via DHN. The configuration of DHN can be changed by remote control of the pipeline valve.

The CHP-VPP operator schedules all units and pipeline valves of DHN within its jurisdiction based on the principle of maximizing renewable energy power generation. Furthermore, the CHP-VPP, as an independent entity, can participate in electricity trading in the electricity market [18]. At a low electricity price, the CHP-VPP chooses to purchase electricity from the grid, reducing internal power generation to reduce costs; on the contrary, it sells electricity to the grid to increase the revenue.

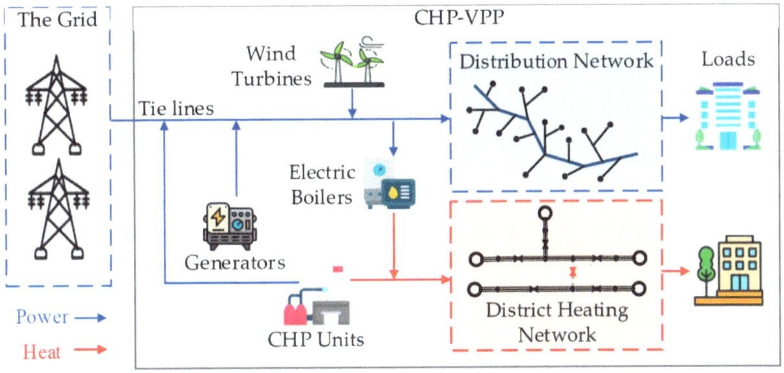

Figure 1. The structure of CHP-VPP.

2.1. Objective Function

The model proposed in our study purposes to minimize the operation cost of the CHP-VPP during the scheduling period.

$$min\ C = C^s + C^f + C^e \tag{1}$$

where C^s and C^f are the start-up costs and fuel costs of the units in the CHP-VPP, respectively; C^e is the cost of interacting with the power grid.

2.1.1. Start-Up Costs

Each start-up of the units requires a fixed cost, as follows:

$$C^s = \sum_{i \in I_G \cup I_{CHP}} \sum_{t=1}^{24} \lambda_i^s y_t^i + \sum_{j \in I_{EB}} \sum_{t=1}^{24} \lambda_j^s y_t^j \tag{2}$$

where λ_i^s/λ_j^s are the start-up cost coefficients of unit i except EB/EB j; y_t^i/y_t^j are binary variables, which represent the starting state of unit i except EB/EB j; and I_G, I_{CHP} and I_{EB} represent the set of generators, CHP units and EBs, respectively.

2.1.2. Fuel Costs

The fuel costs are calculated according to the output power of each unit, as shown below:

$$C^f = \sum_{i \in I_G} \sum_{t=1}^{24} \lambda_i^f p_t^i + \sum_{i \in I_{CHP}} \sum_{t=1}^{24} (\lambda_i^e p_t^i + \lambda_i^h h_t^i) \tag{3}$$

where λ_i^f is the fuel cost coefficient of generator i; λ_i^e and λ_i^h are the electric and heat cost coefficients of CHP i; and p_t^i and h_t^i are the outputs of active power and heat of unit i, respectively.

2.1.3. Costs of Interacting with the Power Grid

The CHP-VPP chooses to purchase or sell electricity at each time period according to the electricity price.

$$C^e = \sum_{t=1}^{24} (\lambda_t^{sell} x_t^{sell} + \lambda_t^{pur} x_t^{pur}) p_t^e \tag{4}$$

where p_t^e is the power that interacts with the grid. When $p_t^e > 0$, it denotes that the CHP-VPP acquires electricity from the grid; on the contrary, when $p_t^e < 0$, it means the CHP-VPP sells electricity to the grid. λ_t^{sell} and λ_t^{pur} indicate the price of selling and purchasing electricity at time t. x_t^{sell} and x_t^{pur} are binary variables that indicate the state of CHP selling and purchasing electricity at time t.

2.2. Operation Constraints of PDS

In our study, a DistFlow power flow equation [19] is utilized to construct the active power flow (5), reactive power flow (6) and voltage drop (7) equations of the branch in PDN.

$$\sum_{d\in\pi(j)} P_t^{jd} + \hat{p}_t^{WT,j} + p_t^{CHP,j} + p_t^{G,j} - p_t^{EB,j} - P_t^{EL,j} - P_t^{ij} = 0 \tag{5}$$

$$\sum_{d\in\pi(j)} Q_t^{jd} + \hat{q}_t^{WT,j} + q_t^{CHP,j} + q_t^{G,j} - q_t^{EB,j} - Q_t^{EL,j} - Q_t^{ij} = 0 \tag{6}$$

$$V_t^i - V_t^j = \left(P_t^{ij} r^{ij} + Q_t^{ij} x^{ij}\right) / V_t^{base}, \forall t \tag{7}$$

where P_t^{ij}/Q_t^{ij} are the active/reactive power, respectively; V_t^j/V_t^{base} are the voltage/voltage reference values; r^{ij}, x^{ij} are the impedance of branch (i,j); $p_t^{CHP,j}, p_t^{G,j}, p_t^{EB,j}, \hat{p}_t^{WT,j}$ represent the active power of the CHP unit/generator/EB/WT connected to node j; $q_t^{CHP,j}, q_t^{G,j}, q_t^{EB,j}$, $\hat{q}_t^{WT,j}$ represent the output reactive power of the CHP unit/generator/EB/WT connected to node j; $P_t^{EL,j}/Q_t^{EL,j}$ are the active/reactive power of electric load; and $\pi(j)$ is the set of children nodes of node j.

Constraint (8) defines the capacity limit of the branch and constraint (9) describes the limit of the node voltage.

$$P_t^{ij,min} \le P_t^{ij} \le P_t^{ij,max}, Q_t^{ij,min} \le Q_t^{ij} \le Q_t^{ij,max}, \forall t \tag{8}$$

$$V_t^{j,min} \le V_t^j \le V_t^{j,max}, \forall t \tag{9}$$

where $P_t^{ij,min}/Q_t^{ij,min}$ are the minimum active/reactive power of branch (i,j), respectively; $P_t^{ij,max}/Q_t^{ij,max}$ are the maximum active/reactive power of branch (i,j); and $V_t^{j,min}/V_t^{j,max}$ are the minimum/maximum voltage value of node j.

Constraint (10) is the power balance equation of the whole CHP-VPP.

$$p_t^G + \hat{p}_t^{WT} + p_t^{CHP} + p_t^e = P_t^{EL} + p_t^{EB}, \forall t \tag{10}$$

Constraints (11) and (12) regulate the start-up operation of all devices. Constraint (13) defines the output range of all units, and the ramping limits are restrained by constraint (14).

$$x_t^i - x_{t-1}^i \le y_t^i, x_t^j - x_{t-1}^j \le y_t^j, \forall t, \forall i \in I_G \cup I_{CHP}, \forall j \in \cup I_{EB} \tag{11}$$

$$\sum_{t=1}^{24} y_t^i \le N^i, \sum_{t=1}^{24} y_t^j \le N^j, \forall t, \forall i \in I_G \cup I_{CHP}, \forall j \in I_{EB} \tag{12}$$

$$x_t^i P^{i,min} \le p_t^i \le x_t^i P^{i.max}, x_t^j P^{j,min} \le p_t^j \le x_t^j P^{j.max}, \forall t, \forall i \in I_G \cup I_{CHP}, \forall j \in \cup I_{EB} \tag{13}$$

$$-\Delta^{i,min} \le p_t^i - p_{t-1}^i \le \Delta^{i,max}, -\Delta^{j,min} \le p_t^j - p_{t-1}^j \le \Delta^{j,max}, \forall t, \forall i \in I_G \cup I_{CHP}, \forall j \in \cup I_{EB} \tag{14}$$

where x_t^i/x_t^j are binary variables and 1 indicates that the unit is in operation; N^i/N^j are the maximum starting times of generator i and CHP i/EB j during the scheduling period; $P^{i.max}/P^{i,min}$ are the maximum/minimum output power of the unit i except EB; $P^{j.max}/P^{j,min}$ are the maximum/minimum output power of EB j; p_t^j is the output of active power of EB j; $\Delta^{i.max}/\Delta^{i,min}$ are the maximum/minimum ramping values of the unit I except EB; and $\Delta^{j.max}/\Delta^{j,min}$ are the maximum/minimum ramping values of EB j.

In addition, the interaction with the grid is constrained by (15), which means that the CHP-VPP cannot buy and sell electricity in one period of time.

$$x_t^{sell} + x_t^{pur} \leq 1, -Mx_t^{pur} \leq p_t^e \leq Mx_t^{sell}, \forall t \tag{15}$$

where M is a very large, positive, real number.

2.3. Operation Constraints of Reconfigurable DHS

The traditional, exact DHS model contains a hydraulic–thermodynamic equation with nonlinear terms and is only applicable to the case where the mass flow direction does not change. Therefore, when solving the CHP-VPP optimal dispatch problem considering DHN reconfiguration, it is essential to convert the exact DHS model into the energy flow model [20]. The core of its transformation is to take the available heat quantity $h_{p,t} = cm_{p,t}\left(\tau_{p,t}^S - \tau_{p,t}^R\right)$ in the pipeline as the decision variable and approximate the heat loss.

Similar to the power balance constraint, constraint (16) is the heat balance equation.

$$\sum_{p \in I_j^{P+}} h_{p,t}^{Out} + \sum_{s \in I_j^{HS}} h_{s,t}^{HS} = \sum_{l \in I_j^{HL}} h_{l,t}^{HL} + \sum_{p \in I_j^{P-}} h_{p,t}^{In} \tag{16}$$

where I_j^{HS}, I_j^{HL} represent the sets of heat station and heat load connected to node j; I_j^{P+} / I_j^{P-} represent the set of pipelines flowing from/to node j; $h_{p,t}^{In}$ / $h_{p,t}^{Out}$ are the inlet/outlet heat flow of pipeline p; $m_{p,t}$ is the mass flow of pipeline p; and $\tau_{p,t}^S / \tau_{p,t}^R$ is the mass flow temperature in pipeline p of the supply/return network.

Constraint (17) indicates that the heat of the heat station comes from the CHP unit and EB.

$$\sum_{i \in I_s^{CHP}} h_t^i + \sum_{j \in I_s^{EB}} h_t^j = h_{s,t}^{HS}, \forall s \in I_{HS}, \forall t \tag{17}$$

where I_s^{EB}, I_s^{CHP} represent the set of EBs and CHP units connected to the heat station s and h_t^j is the heat output of EB j.

Constraints (18) and (19) are limits on the CHP unit, including the limitation of combination factor $\lambda_t^{i,k}$ related to the output of CHP unit i and the limitation of the CHP unit output.

$$0 \leq \lambda_t^{i,k} \leq 1, \sum \lambda_t^{i,k} = x_t^i, \forall k \in OZ^i, \forall t, \forall i \in I_{CHP} \tag{18}$$

$$p_t^{CHP} = \sum_{k \in OZ^i} \lambda_t^{i,k} P^{i,k}, h_t^{CHP} = \sum_{k \in OZ^i} \lambda_t^{i,k} H^{i,k}, \forall t, \forall i \in I_{CHP} \tag{19}$$

where OZ is the set of extreme points in the operation feasible area of CHP unit and $P^{i,k}/H^{i,k}$ are electric/heat power output corresponding to the k-th extreme point, respectively.

Constraint (20) represents the relationship between heat and power output of EB.

$$h_t^{EB} = \lambda^{EB} p_t^{EB}, \forall t \tag{20}$$

where λ^{EB} is the heat–electric coefficient of EB unit.

Constraints (21) and (22) define the meaning of the heat loss and its calculation formula. The heat quality of the pipeline is limited by constraint (23) due to the user's requirements for the heating temperature.

$$h_{p,t}^{Out} = h_{p,t}^{In} - x_t^p h_{p,t}^{Loss}, \forall p \in I_P, \forall t \tag{21}$$

$$h_{p,t}^{Loss} = \frac{\lambda L}{A\rho}\left(\tau_{p,t}^{In,S,min} + \tau_{p,t}^{In,R,min} - 2\tau_t^{Am}\right), \forall p \in I_P, \forall t \tag{22}$$

$$-x_t^p h_{p,t}^{max} \leq h_{p,t}^{Out}, h_{p,t}^{In} \leq x_t^p h_{p,t}^{max} \tag{23}$$

where x_t^p represents the state of the pipeline p, $h_{p,t}^{Loss}$ is the heat loss of pipeline p, λ is the heat conductivity of the pipeline p, ρ/c are the density/heat capacity of water, L is the length of pipeline p, A is the cross-sectional area of pipeline p, $\tau_{p,t}^{In,S,min}/\tau_{p,t}^{In,R,min}$ represents the minimum inlet temperature of pipeline p in the supply/return network, τ_t^{Am} is the ambient temperature, and $h_{p,t}^{max}$ denotes the upper limit of heat quantity in pipeline p.

Constraint (24) indicates the boundary restraint of unit heat output.

$$x_t^i H^{i,min} \le h_t^i \le x_t^i H^{i,max}, x_t^j H^{j,min} \le h_t^j \le x_t^j H^{j,max}, \forall t, \forall i \in I_{CHP}, \forall j \in \cup I_{EB} \quad (24)$$

where $H^{i,max}/H^{i,min}$ are the upper and lower limits of the heat output of CHP i and $H^{j,max}/H^{j,min}$ are the upper and lower limits of the heat output of EB j.

The DHN logic constraints (25) and (26) indicate that the total number of pipeline openings before and after reconfiguration cannot change.

$$x_t^p - x_{t-1}^p = a_t^p - b_t^p, \quad \sum_{r=\max\{1,t-T^p+1\}}^{t} a_r^p \le x_t^p, \quad \sum_{r=\max\{1,t-T^p+1\}}^{t} b_r^p \le 1 - x_t^p \quad (25)$$

$$\sum x_t^p = \sum \tilde{x}_t^p \quad (26)$$

where a_t^p/b_t^p are binary variables representing the opening/closing state of pipeline p, respectively; T^p is the minimum interval time for pipeline operation; and $\tilde{x}_t^p$ is the number of opened pipelines.

2.4. Two-Stage Robust Optimization Model

The above model does not consider the uncertainty of wind farm. Real-time wind output cannot be accurately predicted. Therefore, this paper constructs an adaptive robust optimization (ARO) model considering the uncertainty of wind power, which can be expressed as the following compact form:

$$\begin{aligned} &\min_{x} \boldsymbol{a}^T \boldsymbol{x} \max_{u \in \Theta^u} \min_{y} \boldsymbol{b}^T \boldsymbol{y} \\ &s.t. \begin{cases} \boldsymbol{Ax} = 0, \boldsymbol{Bx} \le \boldsymbol{m} \\ \boldsymbol{Cy} = 0, \boldsymbol{Dy} \le \boldsymbol{n}, \boldsymbol{Ex} + \boldsymbol{Fu} + \boldsymbol{Gy} \le \boldsymbol{q} \end{cases} \end{aligned} \quad (27)$$

where $\boldsymbol{x}$ represents the first-stage variables, containing the start-up states of all devices and the on/off state of valves; $\boldsymbol{u}$ denotes the uncertainty variables, specifically referring to the randomness of the wind output value; $\boldsymbol{y}$ represents the second-stage decision variables, including the electric power and heat output of all devices under confirmed wind generation.

The uncertainty set of wind output Θ^u is as follows, where a robust control coefficient Γ^{WT} is introduced to avoid an over-conservative decision:

$$\begin{aligned} \Theta^u = \{\boldsymbol{u} | \tilde{p}_t^{WT} = \hat{p}_t^{WT} + (\overline{p}_t^{WT} - \hat{p}_t^{WT}) \cdot b_t^+ - (\hat{p}_t^{WT} - \underline{p}_t^{WT}) \cdot b_t^-, \\ b_t^+, b_t^- \in [0,1], \sum_{t=1}^{24} (b_t^+ + b_t^-) \le \Gamma^{WT}\} \end{aligned} \quad (28)$$

where $\tilde{p}_t^{WT}/\hat{p}_t^{WT}$ are the actual/predicted values of wind power and $\overline{p}_t^{WT}/\underline{p}_t^{WT}$ are the upper and lower limits of wind power.

The value of Γ^{WT} is between 0 and 24, which indicates the maximum number of periods in which the actual wind power generation can fluctuate relative to the predicted value.

3. Solution Methodology

The above ARO model commonly uses C&CG algorithm to effectively deal with the global optimal solution [21]. The algorithm is a method that divides the primary problem into one master problem and one subproblem for iterative solution. The subproblem is used to derive the worst scenario in the uncertainty set; then, the corresponding variables and constraints are introduced to the master problem according to the scenario.

3.1. Master Problem of C&CG

The master problem solves the variables that satisfy the first-stage optimization object according to the worst scenario $\boldsymbol{u}_k^*$ generated by the subproblem.

$$\begin{aligned} &\min \boldsymbol{a}^T\boldsymbol{x} + \eta \\ &s.t.\ \eta \geq \boldsymbol{b}^T\boldsymbol{y}_k, \forall k = 1, \ldots, n \\ &\quad \boldsymbol{A}\boldsymbol{x} = 0, \boldsymbol{B}\boldsymbol{x} \leq \boldsymbol{m} \\ &\quad \boldsymbol{C}\boldsymbol{y}_k = 0, \boldsymbol{D}\boldsymbol{y}_k \leq \boldsymbol{n}, \boldsymbol{E}\boldsymbol{x} + \boldsymbol{F}\boldsymbol{u}_k^* + \boldsymbol{G}\boldsymbol{y}_k \leq \boldsymbol{q} \end{aligned} \tag{29}$$

3.2. Subproblem of C&CG

According to the optimal first-stage decision variables, the subproblem solves the max–min model to obtain the worst scenario $\boldsymbol{u}_k^*$ and the optimal object value.

$$\begin{aligned} \eta(\boldsymbol{x}^*, \boldsymbol{u}, \boldsymbol{y}) = \max_{\boldsymbol{u} \in \Theta^u} \min_{\boldsymbol{y}} &\ \boldsymbol{b}^T\boldsymbol{y} \\ s.t.&\ \boldsymbol{C}\boldsymbol{y} = 0 \\ &\ \boldsymbol{D}\boldsymbol{y} \leq \boldsymbol{n} \\ &\ \boldsymbol{E}\boldsymbol{x} + \boldsymbol{F}\boldsymbol{u} + \boldsymbol{G}\boldsymbol{y} \leq \boldsymbol{q} \end{aligned} \tag{30}$$

In summary, the solving steps of the CHP-VPP two-stage robust optimization are as follows:

(1) Initialize the iteration number $n = 0$ and set the bounds of the model $UB = +\infty$, $LB = \infty$.

(2) Solve the master problem and obtain the optimal solution $\boldsymbol{x}^*$ and η^*. Update $LB = max\{LB, a^T x^* + \eta^*\}$.

(3) Solve the subproblem with $\boldsymbol{x}^*$; derive the uncertain scenario parameter $\boldsymbol{u}^*$ and the optimal solution $\boldsymbol{y}^*$. Update $UB = min\{UB, \boldsymbol{a}^T\boldsymbol{x}^* + \eta(\boldsymbol{x}^*, \boldsymbol{u}^*, \boldsymbol{y}^*)\}$.

(4) If $UB - LB \leq \varepsilon$, quit the iteration. Otherwise, update the worst scenario in the master problem, then go to (2).

The specific flow chart of the above C&CG algorithm is shown in Figure 2.

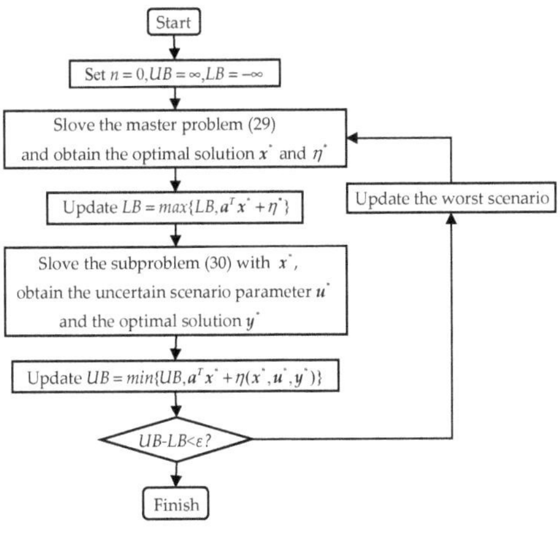

Figure 2. C&CG algorithm flow chart.

4. Case Studies

All tests are resolved using CPLEX interfaced through MATLAB [22].

4.1. Test System Description

Figure 3 depicts the test system selected in this paper. The test system is composed of a 33-node PDS and 8-node DHS, referred to as the P33H8 system. Heating is provided by two heat stations (denoted as HS1 and HS2) in the DHS to satisfy the heat demand. HS1 consists of one EB and one extraction–condensing CHP unit (CHP1), while HS2 is equipped with only one backpressure CHP unit (CHP2). Under normal conditions, the valve v5 is closed and other valves are open, HL3 is supplied by HS2, and HL1 and HL2 are distributed to HS1.

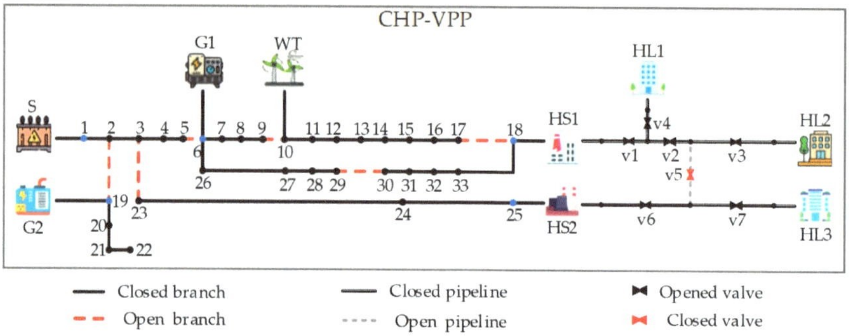

Figure 3. Configuration of the P33H8 system.

4.2. Scenario Settings and Results Analysis

The following three cases are investigated to compare and analyze the results:

Case 1: neither the reconfigurability of DHN nor the uncertainty of wind power are taken into account, i.e., the valve v5 is normally closed, and the output of wind power is its predicted value.

Case 2: the reconfiguration of the DHN is achieved by remote control of the the sectionalizing and tie valves, but the fluctuation of wind power is still ignored.

Case 3: both the reconfigurability of DHN and the uncertainty of wind power are discussed.

The total costs in Case 1 to Case 3 are displayed in Table 1.

Table 1. Costs and revenues in Case 1 to Case 3.

	C^s	C^f	C^e	**Total Cost**
Case 1	6400	72,925.94	7954.03	87,279.97
Case 2	6400	64,158.62	14,295.77	84,854.39
Case 3	6400	64,158.62	17,547.05	88,105.67

4.2.1. Analysis of DHN Reconfiguration

The tie-line power between the CHP-VPP and grid of Case 1 (i.e., ignoring the reconfigurability of DHN) and Case 2 (i.e., considering the DHN reconfiguration) are shown in Figure 4.

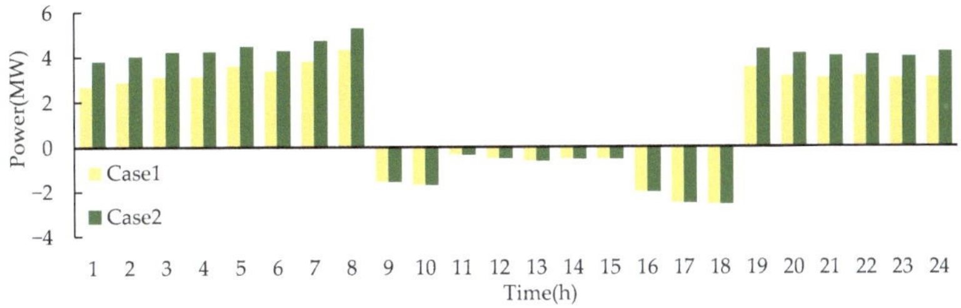

Figure 4. The tie-line power between the CHP-VPP and grid of Case 1 and Case 2.

(1) Ignoring the reconfigurability of DHN

At this time, the value v5 often remains closed, HL1 and HL2 are heated by HS1 and HL3 is heated by HS2. Figure 5a is the output curve of each unit in Case 1.

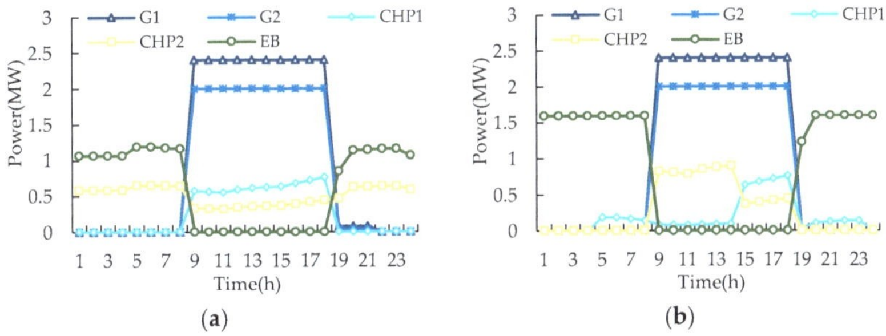

Figure 5. Output curve of each unit: (**a**) Case 1; (**b**) Case 2.

Before 8:00 and after 19:00, HS1 is basically heated by EB, which is due to the lower electricity price during these periods, as it is more economical to purchase electricity from the power grid than to start CHP units for power generation. The output of CHP2 is confirmed by HL3; during 9:00 to 18:00, due to the rise in electricity prices, the output of CHP units is increased to sell more electricity to the grid to earn more revenue. The EB unit is shut down owing to its power consumption properties.

(2) Considering the reconfigurability of DHN

After considering the reconfigurability of DHN, the heat load will be redistributed among the participating heat sources so that the output of each heating unit can be flexibly changed to seek the most economical scheduling plan. As compared with Case 1, the total costs decrease by 2.78% in Case 2. Figure 5b is the output curve of each unit in Case 2. Figure 6 shows different topologies of DHN under reconfiguration operation in Case 2.

During 0:00 to 8:00 and 19:00 to 24:00, the load originally heated by CHP2 in Case 1 can be heated by the less-expensive EB unit since the DHN can be reconfigured. At this time, the value v6 is turned off and all heat loads are supplied by HS1. It can be seen from Figure 5 that the output of CHP2 is reduced, while the EB unit operates at maximum power. When EB cannot meet all the heat load, the CHP1 unit starts to meet the remaining heat load. Therefore, the quantity of electricity purchased from the grid during these periods is higher than in Case 1; from 9:00 to 14:00, the cheaper CHP2 unit provides the heat originally borne by CHP1 in Case 1 to reduce costs; from 15:00 to 18:00, the output of wind power increases significantly. The CHP1 unit with a wider operating range is selected to increase its output and absorb more wind power, while the output of the CHP2 unit is relatively reduced. At this point, the valve v5 is turned off so that HL1 and HL2 are provided by HS1, while HL3 is provided by HS2.

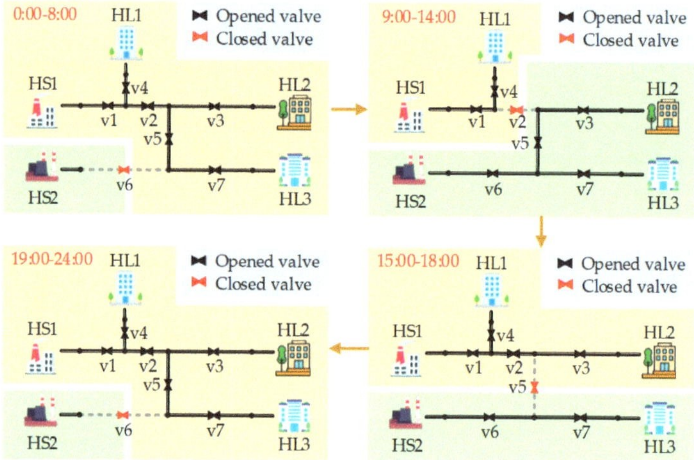

Figure 6. Network topology of different time periods in Case 2.

4.2.2. Analysis of Uncertainty of Wind Power

The wind power value in Case 2 is a certain predicted value, while its output in Case 3 is random and $\Gamma^{WT} = 12$ (i.e., the wind power output can take the maximum or minimum value of the fluctuation interval in 12 periods). Figure 7 displays the wind power output values in Case 2 and Case 3, respectively. Figure 8 shows the tie-line power between the CHP-VPP and grid of Case 2 (i.e., ignoring the uncertainty of wind power) and Case 3 (i.e., considering the uncertainty of wind power).

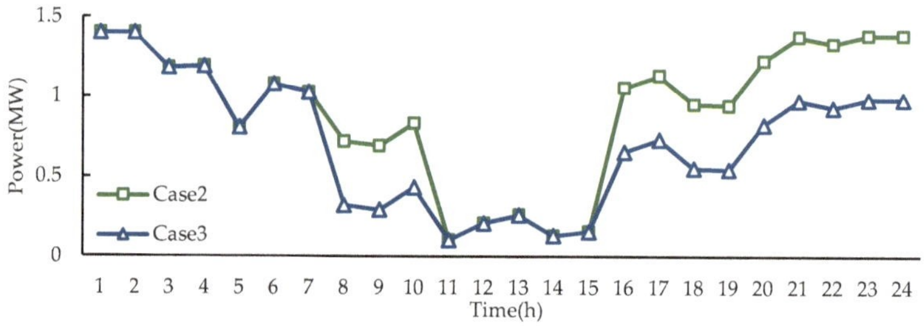

Figure 7. The wind power output values in Case 2 and Case 3.

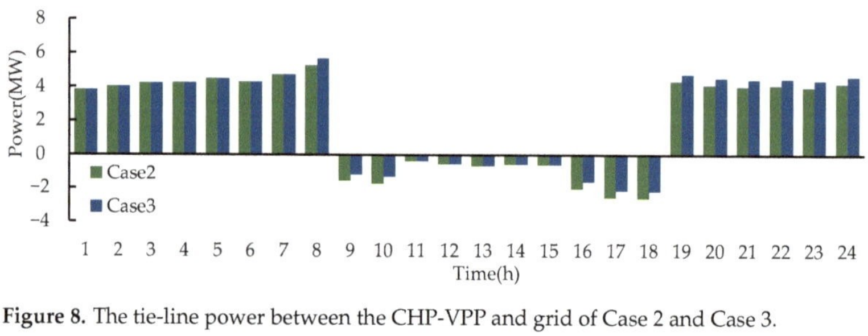

Figure 8. The tie-line power between the CHP-VPP and grid of Case 2 and Case 3.

The worst scenario of the model is that the wind power takes the minimum value of the fluctuation range from Figure 7. Meanwhile, the CHP-VPP needs to buy more electricity from the power grid when the electricity price is low. When the electricity price is high, due to the reduction in wind power output, the quantity of electricity sold by the CHP-VPP to the power grid also decreases accordingly. This leads to an increase in electricity purchase costs, a decrease in electricity sales revenue and, ultimately, a significant increase in the total cost of CHP-VPP.

By adjusting the value of the robustness coefficient, different robustness optimization results can be obtained. The robustness coefficient reflects the ability of the decision-making scheme to resist risks and directly affects the scheduling scheme [23]. Figure 9 shows the total cost of the CHP-VPP under different robustness coefficients.

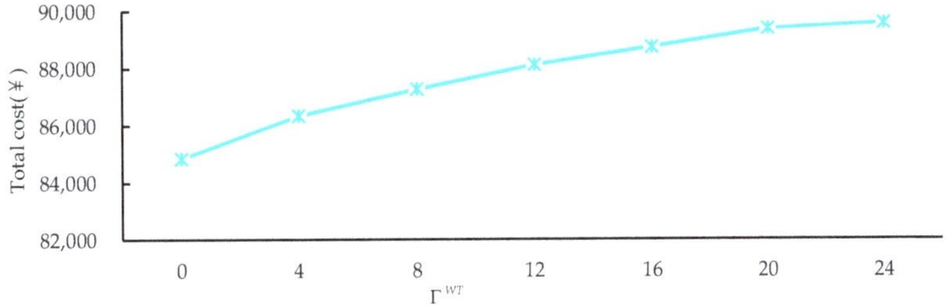

Figure 9. The total cost of the CHP-VPP under different robustness coefficients.

4.2.3. Results of an Actual CHP-VPP in Beijing, China

The actual CHP-VPP in Beijing includes a 7-node PDS and 11-node DHS [24], equipped with three CHP units, one gas boiler and three WTs. Among them, HS2 (connected to node 2) and HS3 (connected to node 6) are provided by one CHP unit, respectively, and HS1 (connected to node 1) includes a CHP unit and a gas boiler. The heat load is connected at nodes 7, 8, 9, 10 and 11. The total cost of Case 1 and Case 2 in the system is shown in Table 2.

Table 2. Costs and revenues in Case 1 and Case 2 in the actual CHP-VPP.

	C^s	C^f	C^e	**Total Cost**
Case 1	5400	89,103.77	−15,513.90	78,989.87
Case 2	5400	88,698.03	−14,197.65	75,040.38

The total cost of Case 2 is 5.00% lower than that of Case 1 from Table 2, which shows that by redistributing heat load between different heat stations, the heat generation of CHP units can be adjusted more flexibly and the flexibility of the power system can be improved effectively.

5. Conclusions

A CHP-VPP optimal dispatch model considering the reconfigurability of DHN and its optimal operation strategy is proposed in this paper, which significantly reduces the total cost of the park-level CHP-VPP. At the same time, ARO is used to tackle with the randomness of wind power, which enhances the robustness of operation. According to the different risk preferences of operators, different robust coefficients are selected for scheduling schemes.

This paper concentrates on the operation optimization of CHP-VPP. The operation optimization of CHP-VPP considering coordinated reconfiguration of integrated power distribution and heating networks under multi-dimensional uncertainty can be studied in the future.

Author Contributions: Conceptualization, J.W.; Supervision, J.W. and B.W.; Project administration, J.W., G.Y. and S.L.; Methodology, Z.P. and H.G.; Writing—original draft, Z.P. and H.G.; Writing—review & editing, Z.P., H.G. and B.W.; Validation, S.L. and G.Y.; Investigation, S.L. All authors have read and agreed to the published version of the manuscript.

Funding: This research is funded by the Science and Technology Project of Shanxi Electric Power Company No. 52053022000K.

Data Availability Statement: Not applicable.

Conflicts of Interest: The authors declare no conflict of interest.

References

1. Li, Y. New energy utilization in environmental design and realization. *Energy Rep.* **2022**, *8*, 9211–9220.
2. Ghiasi, M. Detailed study, multi-objective optimization, and design of an AC-DC smart microgrid with hybrid renewable energy resources. *Energy* **2019**, *169*, 496–507. [CrossRef]
3. Renewable Capacity Highlights. 2020. Available online: https://www.irena.org/statistics (accessed on 2 June 2020).
4. Jamali, A.; Aghaei, J.; Esmaili, M.; Nikoobakht, A.; Niknam, T.; Shafie-Khah, M.; Catalao, J.P.S. Self-scheduling approach to coordinating wind power producers with energy storage and demand response. *IEEE Trans. Sustain. Energy* **2020**, *11*, 1210–1219. [CrossRef]
5. Li, B.; Ghiasi, M. A new strategy for economic virtual power plant utilization in electricity market considering energy storage effects and ancillary services. *J. Electr. Eng. Technol.* **2021**, *16*, 2863–2874. [CrossRef]
6. Li, J.; Zhu, Y.; Yong, M. Cooperative operation and profit distribution of virtual power plant. *Electr. Power Compon. Syst.* **2023**, *51*, 71–82. [CrossRef]
7. Yang, X.; Niu, D.; Sun, L.; Wang, K.; De, G. Participation of electric vehicles in auxiliary service market to promote renewable energy power consumption: Case study on deep peak load regulation of auxiliary thermal power by electric vehicles. *Energy Sci. Eng.* **2021**, *9*, 1465–1476. [CrossRef]
8. Tan, Z.; Zhong, H.; Xia, Q.; Kang, C.; Wang, X.S.; Tang, H. Estimating the robust P-Q capability of a technical virtual power plant under uncertainties. *IEEE Trans. Power Syst.* **2020**, *35*, 4285–4296. [CrossRef]
9. Kardakos, E.G.; Simoglou, C.K.; Bakirtzis, A.G. Optimal offering strategy of a virtual power plant: A stochastic bi-level approach. *IEEE Trans. Smart Grid* **2016**, *7*, 794–806. [CrossRef]
10. Wang, S.; Wu, W. Aggregate flexibility of virtual power plants with temporal coupling constraints. *IEEE Trans. Smart Grid* **2021**, *12*, 5043–5051. [CrossRef]
11. Mohsen Hosseini, S.; Carli, R.; Jantzen, J.; Dotoli, M. Multi-block ADMM approach for decentralized demand response of energy communities with flexible loads and shared energy storage system. In Proceedings of the 2022 30th Mediterranean Conference on Control and Automation (MED), Vouliagmeni, Greece, 28 June–1 July 2022; pp. 67–72.
12. Ancona, M.A.; Baldi, F.; Branchini, L.; De Pascale, A.; Gianaroli, F.; Melino, F.; Ricci, M. Comparative analysis of renewable energy community designs for district heating networks: Case study of corticella (Italy). *Energies* **2022**, *15*, 5248. [CrossRef]
13. Daniel, F.; Erik, M.; Russell, M.; Ute, K.; Wolf, F. On the economic potential for electric load management in the German residential heating sector—An optimizing energy system model approach. *Energy* **2014**, *71*, 263–276.
14. Fang, F.; Yu, S.; Liu, M. An improved Shapley value-based profit allocation method for CHP-VPP. *Energy* **2020**, *213*, 118805. [CrossRef]
15. Kordestani, M.; Zanj, A.M.; Orchard, E.; Saif, M. A modular fault diagnosis and prognosis method for hydro-control valve system based on redundancy in multisensory data information. *IEEE Trans. Reliab.* **2019**, *68*, 330–341. [CrossRef]
16. Du, Y.; Xue, Y.; Wu, W.; Shahidehpour, M.; Shen, X.; Wang, B.; Sun, H. Coordinated planning of integrated electric and heating system considering the optimal reconfiguration of district heating network. *IEEE Trans. Power Syst.* **2023**, 1–14. [CrossRef]
17. Wang, K.; Xue, Y.; Guo, Q.; Shahidehpour, M.; Zhou, Q.; Wang, B.; Sun, H. A coordinated reconfiguration strategy for multi-stage resilience enhancement in integrated power distribution and heating networks. *IEEE Trans. Smart Grid* **2022**, *14*, 2709–2722. [CrossRef]
18. Zhao, H.; Wang, B.; Pan, Z.; Sun, H.; Guo, Q.; Xue, Y. Aggregating additional flexibility from quick-start devices for multi-energy virtual power plants. *IEEE Trans. Sustain. Energy* **2021**, *12*, 646–658. [CrossRef]
19. Yeh, H.-G.; Gayme, D.F.; Low, S.H. Adaptive VAR control for distribution circuits with photovoltaic generators. *IEEE Trans. Power Syst.* **2012**, *27*, 1656–1663. [CrossRef]
20. Xue, Y.; Shahidehpour, M.; Pan, Z.; Wang, B.; Zhou, Q.; Guo, Q.; Sun, H. Reconfiguration of district heating network for operational flexibility enhancement in power system unit commitment. *IEEE Trans. Sustain. Energy* **2021**, *12*, 1161–1173. [CrossRef]
21. Zeng, B.; Zhao, L. Solving two-stage robust optimization problems using a column-and-constraint generation method. *Oper. Res. Lett.* **2013**, *41*, 457–461. [CrossRef]
22. Hong, T.; Zhao, D.; Zhang, Y.; Cui, B.; Tian, Y. Optimal voltage reference for droop-based DERs in distribution systems. *IEEE Trans. Smart Grid* **2020**, *11*, 2357–2366. [CrossRef]
23. Baringo, A.; Baringo, L.; Arroyo, J.M. Day-ahead self-scheduling of a virtual power plant in energy and reserve electricity markets under uncertainty. *IEEE Trans. Power Syst.* **2019**, *34*, 1881–1894. [CrossRef]
24. Qin, X.; Shen, X.; Guo, Y.; Pan, Z.; Guo, Q.; Sun, H. Combined electric and heat system testbeds for power flow analysis and economic dispatch. *CSEE J. Power Energy Syst.* **2021**, *7*, 34–44.

MDPI AG
Grosspeteranlage 5
4052 Basel
Switzerland
Tel.: +41 61 683 77 34

Electronics Editorial Office
E-mail: electronics@mdpi.com
www.mdpi.com/journal/electronics

www.ingramcontent.com/pod-product-compliance
Lightning Source LLC
Chambersburg PA
CBHW041347210926
51850CB00013B/3015
9783725831906